VOLUME TWO HUNDRED AND TWENTY SEVEN

ADVANCES IN IMAGING AND ELECTRON PHYSICS

Coherent Electron Microscopy: Designing Faster and Brighter Electron Sources

VOLUME TWO HUNDRED AND TWENTY SEVEN

ADVANCES IN IMAGING AND ELECTRON PHYSICS

Coherent Electron Microscopy: Designing Faster and Brighter Electron Sources

FLORENT HOUDELLIER

Edited by

MARTIN HŸTCH
CEMES-CNRS
Toulouse, France

PETER W. HAWKES
CEMES-CNRS
Toulouse, France

ELSEVIER

ACADEMIC PRESS
An imprint of Elsevier

Cover photo credit: "Design for a magnetic-field superimposed ultrafast cold-field emission gun" by Florent Houdellier.

Academic Press is an imprint of Elsevier
125 London Wall, London EC2Y 5AS, United Kingdom
525 B Street, Suite 1650, San Diego, CA 92101, United States
50 Hampshire Street, 5th Floor, Cambridge, MA 02139, United States
The Boulevard, Langford Lane, Kidlington, Oxford OX5 1GB, United Kingdom

ISBN: 978-0-443-19324-8
ISSN: 1076-5670

For information on all Academic Press publications
visit our website at https://www.elsevier.com/books-and-journals

Publisher: Zoe Kruze
Acquisitions Editor: Jason Mitchell
Developmental Editor: Jhon Michael Peñano
Production Project Manager: James Selvam
Cover Designer: Greg Harris

Typeset by STRAIVE, India

Dedication

To my daughter Lalie, our shooting star …

Contents

Acronyms

ADF annular dark field
BEM boundary element method
BF bright field
BFEH bright field electron holography
BFP back focal plane
BPC condensor biprism
CAD computer-aided design
CBED convergent beam electron diffraction
CCD charge-coupled device
CcNT carbon cone nanotube
CFEG cold field emission gun
CNT carbon nanotube
DED direct electron detector
DFEH dark field electron holography
DLS damped least square
DQE detector quantum efficiency
DTEM dynamic transmission electron microscope
EEGS electron energy gain spectroscopy
EELS electron energy loss spectroscopy
EOD electron optics design
EQBD electrostatic quadrupole beam deflector
FDM finite difference method
FEM finite element method
FE-TEM field emission transmission electron microscope
FFT fast Fourier transform
FIB focused ion beam
FWHM full width at half maximum
GEO geometrical electron optics
GIF Gatan imaging filter
GPA geometrical phase analysis
HAADF high angle annular dark field
HHT Hitachi high-technologies
HOLZ high-order Laue zone
HREM high resolution electron microscopy
MCFEG preaccelerated magnetic lens CFEG
MEBS Munro's electron beam software
MTF modulation transfer function
NEPP number of electrons per pulse
SEA spectrometer entrance aperture
SEM scanning electron microscope
SFEG Schottky field emission gun
SHG second harmonic generation
SIMS secondary ion mass spectrometry

STEM scanning transmission electron microscope
TEM transmission electron microscope
UHV ultra high vacuum
UTEM ultrafast transmission electron microscope
ZAP zone axis pattern
ZLP zero loss peak
ZOLZ zero-order Laue zone

Symbols

λ electron wavelength

m_e electron mass

e magnitude of the electron charge

α electron trajectory angle relative to the reference particle (optic axis) in the plane (xOz)

β electron trajectory angle relative to the reference particle (optic axis) in the plane (yOz)

V_0 acceleration voltage

E_0 kinetic energy of the particle

$\vec{r}$ real space position vector

$\vec{q}$ general reciprocal space vector

$\vec{g}_k$ particle kinetic momentum

$\vec{p}$ particle canonical momentum

n refractive index

V point characteristic function defined in conventional optics

W action

L Lagrangian function

H Hamiltonian function

E total energy of the particle

S point characteristic function defined in charged particle optics

Ψ wave function

φ electron wave phase

B brightness

B_r reduced brightness

γ_{total} total complex degree of coherence

γ_s spatial complex degree of coherence

γ_t temporal complex degree of coherence

Φ_j Bloch wave function j

γ_j Anpassungen of the Bloch wave j

$\vec{k}^j$ wave vector of the Bloch wave function j

C^j excitation amplitude of the Bloch wave function j

$\vec{g}$ lattice plane reciprocal space vector or diffraction vector

$\vec{u}$ displacement field vector

$\vec{E}$ electrostatic field

$U(\vec{r})$ electrostatic potential

$U_m(z, s)$ component m of the electrostatic potential series expansion

$u_m(z)$ axial distribution of the component U_m

$\vec{A}$ magnetic vector potential

$\vec{H}$ magnetic field strength

$\vec{B}$ magnetic field flux density

F general variational function

C_s spherical aberration coefficient of rotationally symmetric lens

$\boldsymbol{C_c}$ chromatic aberration coefficient of rotationally symmetric lens
$\boldsymbol{E_F}$ energy of the Fermi level
$\boldsymbol{\Phi}$ material exit work function
$\boldsymbol{V_1}$ extraction voltage
$\boldsymbol{V_2}$ focusing voltage
$\boldsymbol{N_e}$ number of electrons per pulse
$\boldsymbol{R}$ Butler lens ratio
$\boldsymbol{d_v}$ FE virtual source size

Preface

The development of new instrumentation is often a long and arduous task, not less so for electron microscopy. A brilliant design can be held back at the prototyping stage by any number of mundane issues, from mechanical to electrical. Even if successful, there is no guarantee that manufacturers will develop the technology for practical or commercial reasons, limiting the availability of the innovation to the wider community. The development of aberration correction is a perfect illustration of these aspects, and we refer interested readers to Volume 153 of *AIEP* and especially to the chapters by Haider, Rose, and Krivanek. Examples of instrumentation currently under development can also be found in the Proceedings of the International Conference on Charged Particle Optics (CPO-10) in Volume 212.

We are therefore proud to introduce the work of Florent Houdellier in this volume of *AIEP*, whose instrumental developments have been exemplary in many ways. He has notably contributed to the development of ultrafast electron microscopy, pioneered by Bostanjoglo (see Volume 110 of *AIEP*), for which he shared the Ruska Prize; the first electron microscope with a laser-driven cold-field emission gun was presented with coworkers Arnaud Arbouet and Giuseppe-Mario Caruso in Volume 207. Here, he explains in detail the motivation for his work and sets out his very recent designs for high-brightness cold-field emission guns. His objectives are presented at greater length in the text appended to this preface.

In *AIEP*, we are always happy for authors to share their ideas and future projects with a wider audience and give their personal opinion on opportunities in their field of expertise.

Martin Hÿtch and Peter W. Hawkes
Toulouse

This manuscript is not only a review of my research concerning methodological and instrumental developments in coherent transmission electron microscopy (TEM), but also a scientific project. In particular, I will outline the original concepts for the new coherent sources that I intend to develop. Combining a review with a project is typically required of all French researchers at some point in their careers as part of a qualification called the *Habilitation à Diriger les Recherches* (HDR). But from the beginning, I wanted to write with the goal of publication in *Advances in*

Imaging and Electron Physics and, thus, share the strategy we want to pursue in Toulouse toward the development of new coherent electron sources with the electron optical community.

During my PhD, postdoctoral studies, and the first years of my work as a member of the Centre National de la Recherche Scientifique (CNRS) in the Centre d'Elaboration de Matériaux et d'Etudes Structurales (CEMES), I was mainly trying to develop new methods to study materials at the nanoscale using TEM. I insisted on coherence-based approaches because of their ability to provide highly quantitative information. Mapping crystal strain was the guiding application of these developments. I realized we had to delve deeper into the understanding of the instrument itself and develop key components not provided by TEM manufacturers, with which the sensitivity of existing techniques would be improved and new ones implemented. These developments are reviewed in the first two chapters. Research is obviously a team effort, and all the results that follow were made possible by many discussions with my CNRS colleagues and by the work of the two PhD students, Giuseppe-Mario Caruso and Julien Dupuy, whom I had the opportunity of guiding as a codirector.

In the third and final chapter, I detail my new project, which entails using modern electron optics to design the coherent electron sources of the future. Recently, the development of ingenious optical configurations has indeed deeply transformed the very structure of TEMs and their use. I am thinking, of course, of aberration correctors, which have made a huge contribution to our community. However, for decades, the optics of such coherent sources as field-emission sources has remained relatively unchanged. The main reasons for this were the absence of any pressing need for improvement and the high level of technology required to modify this sensitive part of the microscope. However, the development of coherent sources with new emission properties, such as ultrafast sources, is taking the current optical configurations to their limits, and the need for new optical concepts has become urgent.

In the third chapter, I therefore introduce my project for realizing this objective by first presenting my strategy based on the simulation of complex electron optics systems and then describing three potentially interesting designs that should improve the properties of coherent electron sources. To carry out this project, I rely on the industrial collaboration between CEMES and Hitachi. Developing these complex optical systems will, of course, involve close collaboration with all the CEMES-CNRS support services, notably the mechanical workshop. I am confident that the full realization of at least one of these solutions will benefit the coherent TEM community greatly, especially for ultrafast TEM, and will only be feasible once again thanks to the success of solid teamwork and constructive collaborations.

Florent Houdellier

CHAPTER ONE

Characterization of nanomaterials properties using FE-TEM

Contents

L'homme qui travaille perd un temps précieux.

Miguel de Cervantes

1.1. Principles of physical characterization using a coherent electron beam

1.1.1 The transmission electron microscope: An introduction

The electron is a negatively charged subatomic particle with elemental charge $e = 1.60218 \times 10^{-19}$C and mass $m_e = 9.10938 \times 10^{-31}$kg. J.J. Thomson, a British physicist, made the discovery of the electron in 1897 while experimenting with cathode ray tubes (CRTs) (Smith, 1997). Thomson was able to show that the electrons are negatively charged and to calculate the electron charge/mass ratio using a CRT, deflection plates, and a phosphor screen. According to the particle–wave duality in quantum mechanics, we can define an electron wavelength in the vacuum λ derived

Advances in Imaging and Electron Physics, Volume 227
ISSN 1076-5670
https://doi.org/10.1016/B978-0-443-19324-8.00001-8

using the relationship between λ, the electron kinetic momentum g_k, or the kinetic energy E_0, as proposed by Louis de Broglie in 1924 (De Broglie, 1923):

$$\lambda = \frac{h}{g_k} = \frac{h}{\sqrt{2mE_0}} \tag{1.1}$$

here, m is the electron mass relativistically corrected and h is Planck's constant.

Additionally to Louis de Broglie's theory, electron microscopy is based on additional fundamental discoveries made by Busch (1926). Busch discovered that the magnetic field of a solenoid acts on electrons in the same way as a convex glass lens acts on light rays. Ernst Ruska also begins working on an electron-based microscope that employs electrons rather than photons (Hawkes, 1990). In 1931, he successfully developed the first electron microscope, and for the next 40 years, Ruska's design served as the basis for all transmission electron microscopes (TEMs).

In a conventional TEM, a thin specimen is irradiated with an electron beam of uniform current density. The acceleration voltage of routine instruments is $V_0 = 100$–$300\,\text{kV}$ leading to an electron beam energy in the keV range ($E_0 = eV_0$) (Williams and Barry Carter, 2008). Electrons are emitted in the electron gun whether using intense sources called thermionic source or high brightness sources called Schottky field emission (SFE) or cold field emission (CFE). The latter are used when optimum spatial coherence are needed. TEM working with a field emission source will be defined as "FE-TEM." A three or four stage condenser-lens system allows the illumination aperture and the field of view illuminated on the specimen to be adjusted. Following the Bush discovery, all lenses in a TEM optical column are electromagnetic composed by a coil which generates the magnetic field, and magnetic circuit (yoke + pole piece) which confines the magnetic flux lines between the pole piece gap.

The incident electrons interact with the electrons and nuclei within the sample through Coulomb forces. The incoming electrons are then scattered as the result of these interactions (Hirsch et al., 1977). Fig. 1.1A summarizes various electron scattering effects in a thin, electron transparent sample.

The electron scattering may or may not involve energy loss. While elastic scattering from nearby atoms is typically coherent and experiences no discernible energy loss, inelastic scattering entails a quantifiable quantity of electron loss. The elastic scattering events are further separated into those scattered into relatively low scattering angles or into high angles. Other kind

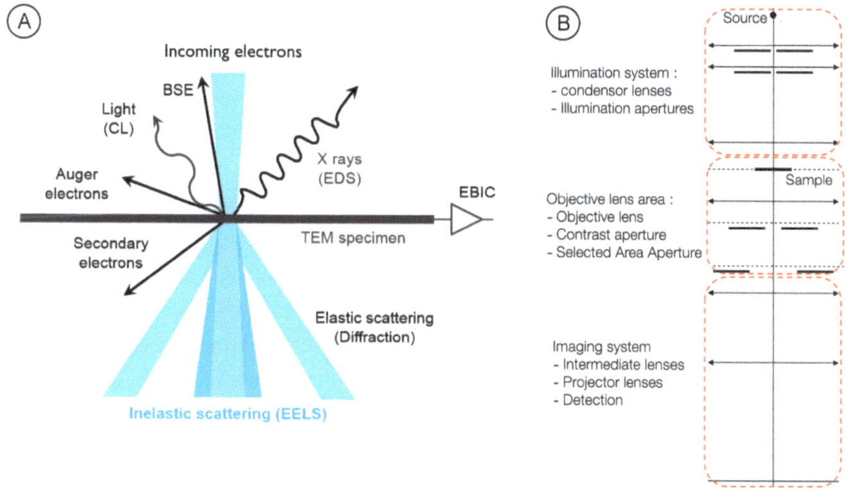

Fig. 1.1 (A) General outline of the signals generated by the interaction of an accelerated electrons beam with a TEM specimen with, among others, catholuminescence (CL), backscattered electrons (BSE), energy dispersive X-ray spectroscopy (EDS), electron energy loss spectroscopy (EELS). (B) General outline of the TEM optic. *Panel (A): From Gatan Inc. https://www.gatan.com/techniques/eels.*

of signals are also generated from the interaction between electrons and the samples such as secondary electrons, Auger electrons, X-ray or visible photons for instance. When the specimen is a crystal, the incoming electrons will then interact with a periodic assembly of atoms. All elastic low angle scattered electrons coming from the interaction with individual atoms, will then interact during the propagation through the crystal. This phenomena will generate constructive direction of interferences called Bragg diffracted beams, at the origin of the electron diffraction pattern observed after the sample (Spence and Zuo, 2013). Information about crystal structure and orientation is provided by the electron diffraction pattern. The possibility of combining electron diffraction and the various imaging modes is the most powerful feature of TEM for the investigation of the crystal lattice and defects in crystalline material (Morniroli, 2002). The diffraction pattern, as well as the electron image, are first formed by the objective lens in two different optical planes located after the sample respectively called diffraction plane and image plane. These planes are then magnified down to the detector by a set of intermediate and projector lenses. A scheme of a conventional TEM is shown in Fig. 1.1B. Before moving on, we make a brief comment on these important concepts introduced within the framework of geometrical electron optics.

Geometrical electron optics describes the motion of electrons in macroscopic electromagnetic fields (electric $\vec{E}$ and/or magnetic $\vec{B}$) by employing the well-established formalism of geometrical light optics (Rose, 2009). When the refraction index does not change significantly over a distance of few wavelengths, which is always the case inside conventional electron optical elements, the influence of electron wave diffraction actually becomes quite negligible.

Due to the analogy with geometrical photon optics, geometrical electron optics conceives the path of an electron as a geometrical line or trajectory. The electron trajectory coordinates can be calculated applying Newton's law of dynamics on the electron which experiences the Lorentz force inside the electromagnetic field. Alternatively, such as using the Fermat principle in conventional optics, electron trajectory could be studied using Hamilton's characteristic function calculated thanks to the Lagrange variational principle of mechanic (Born and Wolf, 1999).

Using both method, the general idea behind the description of the optical properties of a given system will be to express any trajectory coordinates through power expansions, to separate the linear contributions of trajectory starting conditions (i.e., initial positions, angles, energy) from the nonlinear ones (Wollnik, 2014).

Let's consider x and y the cartesian coordinates of an electron crossing an optical plane, located at a position z perpendicular to the optic axis, generally defined by a reference particle. An electron on a given plane z can be fully defined using a set of additional variables: α, β angles relatively to the optic axis, $a = \tan(\alpha)$ along x and $b = \tan(\beta)$ along y, $T = \frac{t - t_{\text{ref}}}{t_{\text{ref}}}$ time of flight difference relative to the reference particle time t_{ref}, $\delta = \frac{E - E_{\text{ref}}}{E_{\text{ref}}}$ energy difference relative to the reference particle energy E_{ref}. A linear form of trajectory equation could be written using the power expansion of the function $x(x_0, y_0, a_0, b_0, T_0, \delta_0, z)$ relative to the initial parameters x_0, y_0, a_0, b_0, T_0, δ_0 in a plane located at $z = z_0$:

$$x(z)_{\text{par}} = (x|x)x_0 + (x|a)a_0 + (x|y)y_0 + (x|b)b_0 + (x|T)T_0 + (x|\delta)\delta_0 \tag{1.2}$$

considering sufficiently small values to the parameters x_0, a_0, y_0, b_0, T_0, δ_0. The linear coefficients of the expansion such as $(x|x) = (\partial x(z)/\partial x_0)_{x_0=0}$ are simply the partial derivatives of the function $x(z)$ with respect to the initial parameters taken in $x_0 = 0$. Using the same notation, nonlinear terms in the power expansions can be written:

$$
\begin{aligned}
x(z)_{ab} = {} & (x|xx)x_0^2 + (x|xa)x_0\,a_0 + (x|aa)a_0^2 + (x|xy)x_0\,y_0 \\
& + (x|a\delta)a_0\,\delta_0 + (x|xxx)x_0^3 + (x|xxa)x_0^2a_0 + (x|xaa)x_0a_0^2 \\
& + (x|aaa)a_0^3 + (x|xyy)x_0y_0^2 + (x|xyb)x_0y_0b_0 + (x|xbb)x_0b_0^2 \\
& + (x|ayy)a_0y_0^2 + (x|ayb)a_0y_0b_0 + (x|abb)a_0b_0^2 + \cdots
\end{aligned}
\tag{1.3}
$$

with $(x|xx) = (1/2!)\,(\partial^2x(z)/\partial x_0^2)_{x_0=0}$ and so on. The general form of the trajectories are described by the sum of the linear and the nonlinear contributions $x(z) = x(z)_{\mathrm{par}} + x(z)_{ab}$. The same expansion can be implemented for the functions $a(z)$, $y(z)$, $b(z)$, $T(z)$, $\delta(z)$ (Wollnik, 2014).

If we consider only the influence of the linear parameters on the trajectory ($x(z) = x(z)_{\mathrm{par}}$), the study is limited to the framework of Gaussian optics or **paraxial optics** as we will call it from now. The linear parameters, such as $(x|x)$, ..., defined the classical paraxial properties of an optical system such as magnification, focal distances, position of gaussian image plane and angular magnification. The nonlinear coefficients of the power expansions are called **aberrations coefficients**. The coefficients containing only geometrical variables (x,y, a,b) in the right part of $(.|.)$ are called geometrical aberrations, while the one containing the energy contribution δ are called chromatic aberrations.

Like the standard Seidel approach in light optics (Gross, 2005), the eikonal method in electron optics can be used to express these nonlinear coefficients as a function of lens field components and paraxial properties, by considering all aberrations as a perturbation of the paraxial rays (Hawkes and Kasper, 2018).

The general idea behind this method will be to develop the lens electromagnetic potentials (electric U and/or magnetic $\vec{A}$) using Fourier series, and each Fourier component can be expanded in power series with respect to spatial coordinates. Using Hamilton's description of an optical problem (detailed more precisely in Section 1.1.2.2) point, mixed or angular characteristic functions (sometimes called point, mixed or angular eikonal functions) could then be also expressed as a function of lens potentials and then also developed in power expansions (Hawkes and Kasper, 2018; Rose, 2009). The paraxial information is encoded in the second-order terms of the characteristic function power expansion. Paraxial solutions acquire extra perturbation terms due to high-order contributions of the characteristic function expansion. The primary (third-order) aberrations of round lenses for instance, are obtained considering the fourth-order term of the characteristic function power expansion as a perturbation of paraxial solutions calculated thanks to the second-order term. Using this method, it becomes possible to express all aberrations coefficients (.|.) encountered in Eq. (1.3) as a function of field potential components expressed in the power expansions, and paraxial properties such as the magnification. This well-known method is detailed in the series "Principle of Electron Optics" (Hawkes and Kasper, 2018) and will be also discussed during the first part of the research project presented in Section 3.2.

In conventional optics an optical plane in which all paraxial rays, originating from the center of the object plane, join in a single point is called an image plane. A plane in which all fundamental rays, originating from the object plane with the same slope, are imaged to a point is called the focal or diffraction plane, previously introduced. In paraxial GEO we introduce fundamental paraxial rays which satisfy the linearized equation of motion, which can be obtained by limiting the calculation of the electron trajectory to the second-order expansion of the Hamilton's characteristic function (see Section 3.2 for details). These equations will then simply have a pair of linearly independent solutions. As for conventional geometrical optics, we can arbitrary define the two independent paraxial solutions as:

1. the marginal ray: the ray that passes from the center of the object, at the maximum aperture of the lens. This ray crosses the optic axis in the image plane.
2. the principal ray: the ray from an off-axis point in the object passing through the center of the aperture stop.

Within a multistage system, like the electron microscope, each plane will be imaged repeatedly as described by the well-known theorem of alternating images (Rose, 2009). In Fig. 1.2 we show the paths of the two fundamental paraxial trajectories (principal and marginal rays) as well as the location of the images and aperture stops (corresponding to the pupil planes) in a TEM.

In Fig. 1.2A the marginal ray starting from the source will define the illuminated field of view on the sample plane located in the crossover of the source principal ray, i.e., the aperture stop plane. This is known as Köhler illumination (Gross, 2005). After crossing the sample the source marginal ray is considered as the sample principal ray. The condition reported in Fig. 1.2A is known as object telescopic condition as the object principal ray starts parallel to the optic axis. The position of the aperture stop is located at the intersection of the object principal ray corresponding to the back focal plane (BFP) of the objective lens under this telescopic illumination. This is the location of the objective aperture, and corresponds to the diffraction plane.

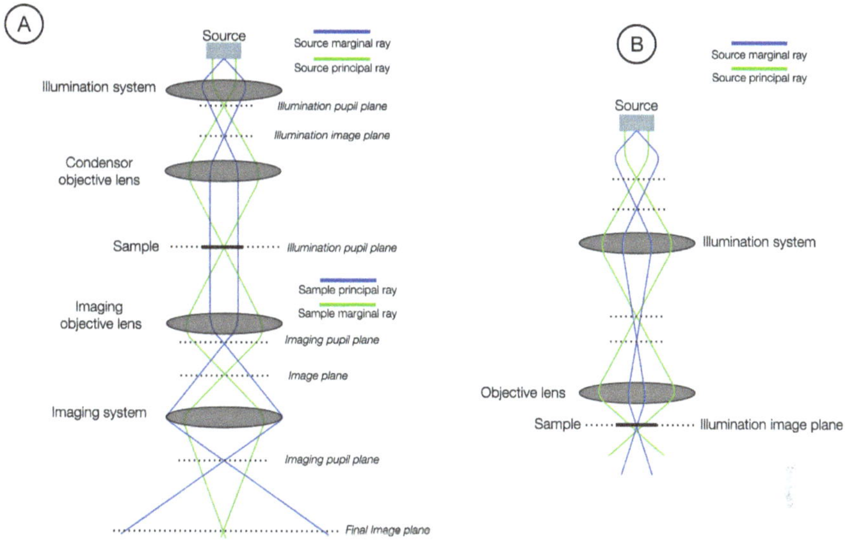

Fig. 1.2 Scheme of the path of the fundamental paraxial trajectories (principal and marginal rays) and location of the images and aperture-stop planes in an electron microscope illustrating the theorem of alternating images. (A) Fundamental paraxial trajectories for a TEM where the sample is located in the illumination exit pupil plane. (B) Fundamental paraxial trajectories for a SEM, or a TEM set in convergent beam focused mode such as STEM/CBED, … where the sample is located in the illumination image plane of the source.

In the case of crystalline specimens, the selection of the transmitted beam (bright field), or a Bragg-reflected beam aligned on-axis (dark field), using the objective aperture in the diffraction plane gives rise to diffraction contrast in the image. This contrast is important for the imaging of crystal defects (Hirsch et al., 1977). In zone axis orientation of the crystal, when many Bragg-diffracted beams also pass through the objective aperture, crystal-structure imaging reveals projections of atomic rows (Williams and Barry Carter, 2008). In the paraxial approximation, the objective aperture size could be defined by the physical size of the electron microscope tube and the resolution could be theoretically limited by the diffraction phenomena of the electron wave such as modern light microscope. But unfortunately, this course of the rays can never be achieved in a real system owing to the unavoidable nonlinear terms in the equation of motion that we briefly described previously.

In most conventional electron microscopes, the spherical aberration of the objective lens (third-order aperture terms as for example $(x|aaa)$ in the aberration expansions (1.3)) determines the achievable resolution of the microscope and the off-axial (Ansys Company, n.d.) (mixed terms: order 2 in aperture and order 1 in position as for instance $(x|aax)$) the field of view of the recorded image. The projector lenses introduce primarily distortion (third-order field terms such as $(x|xxx)$). Hence, to obtain ideal imaging, we must compensate for the spherical aberration and the coma of the objective lens and for the distortion of the projector system. In conventional TEM, distortion can be eliminated by properly exciting the projector lenses. Unfortunately, this is not possible for the spherical aberration due to the Scherzer theorem which states that nonlinear aperture terms $(x|aaa)$, as well as second-order chromatic terms $(x|a\delta)$, cannot be corrected in static electromagnetic lenses with rotationally symmetry (Scherzer, 1936).

The overall geometry of the objective lens polepiece, which is frequently parameterized in terms of the dimensions defining the upper and lower pole-piece faces and bores as well as the gap between the two polepieces, affects both spherical and chromatic aberrations. However, the behavior of both $(x|aaa)$ and $(x|a\delta)$ (and of focal length) can be readily calculated for a given geometry as a function of the lens excitation and specimen position enabling these coefficients to be optimized, subject to constraints imposed by machining tolerances and requirements for specimen movement and tilt (Hawkes, 1982).

Immense effort has been devoted to finding lenses with small aberrations and at the same time to devising aberration correctors. The theoretical proof that the two most important aberrations, the spherical and chromatic aberrations, cannot be eliminated required that several conditions be satisfied and, by relaxing one or another of these conditions, correctors could be designed. These conditions are described in the early work of Scherzer and are referred as the Scherzer conditions (Scherzer, 1936). One condition which could be relaxed concerned the rotationally symmetry of the field. Hence, by using multipolar lenses, it should be possible to find proper optical systems which could correct spherical and/or chromatic aberrations without introducing too much additional aberration such as off-axial coma. Numerous experiments were conducted over the years trying to overcome these intrinsic nonlinear terms of round lenses using clever multipolar design. During the major part of the 20th century, attempts to use aberration correctors to improve the resolution beyond what was possible with the greatest uncorrected electron microscopes failed. Indeed successful aberration-corrected optical systems must accomplish three separate tasks: measure and correct the principal aberrations affecting the resolution, diagnose and eliminate parasitic aberrations mainly coming from mechanical imperfections, and improve the instrumental stabilities. This stage of developments was reached around 1990s for two multipolar design: the quadrupole–octopole corrector and the hexapole corrector (Dellby et al., 2001; Kabius et al., 2002).

The simplest hexapole corrector, called C-COR by the CEOS GmbH company, is made of two long hexapoles lenses located in the conjugate plane of a telescopic round lens doublet (Kabius et al., 2002). This system can be installed after the objective lens. If the coma-free plane of the objective lens is properly imaged in the center of the first hexapole, thanks to a second telescopic doublet of round lenses, then the spherical aberration of the objective lens can be fully compensated without introducing too much off axial coma avoiding a too small corrected field of view. The final arrangement is a 4-f optical system. However this corrector is only a semiaplanatic system because other nonlinear terms remains, limiting the corrected field of view in the range of few 100 nm. Mainly third-order field curvature (terms of order 2 in position and 1 in aperture as for example $(x|xxa)$, ...), field astigmatism (same power as field curvature but mixed x, y for instance $(x|yya)$, ...) and anisotropic coma (imaginary part of complex terms combining $(x|aax)$, $(y|bby)$; only observed in magnetic lenses) remain in the final image. An optical system, which could correct all these geometrical aberrations is called aplanatic and satisfies the Abbe sine condition (Gross, 2005). An aplanatic hexapole arrangement consisting of four identical round lenses forming a 8f-system, three long hexapoles and four short hexapoles, centered symmetrically about the midplane, was manufactured by the CEOS GmbH company and is called B-COR (Müller et al., 2011). Fig. 1.3 reports the arrangement of the elements inside a C-COR and a B-COR spherical-aberration correctors. One example showing the extraordinary improvement in the image quality of a TiAl atomic structure provided by such an original optics system, is also illustrated.

The strength of TEM is that not only provide high-resolution images that contain atomic information but it can also operate with small electron probes focused on the sample in various microanalytical modes (Williams and Barry Carter, 2008). In such an illumination configuration, also reported in Fig. 1.2B, the marginal ray starting from the source will now cross the optic axis on the sample location. The condenser system of a TEM can then be set in Köhler type illumination, where the aperture stop of the illumination system is conjugate to the sample plane, or in focused probe illumination where the image plane of the condenser lenses is now conjugate to the sample plane. The switch from one illumination to another is continuous

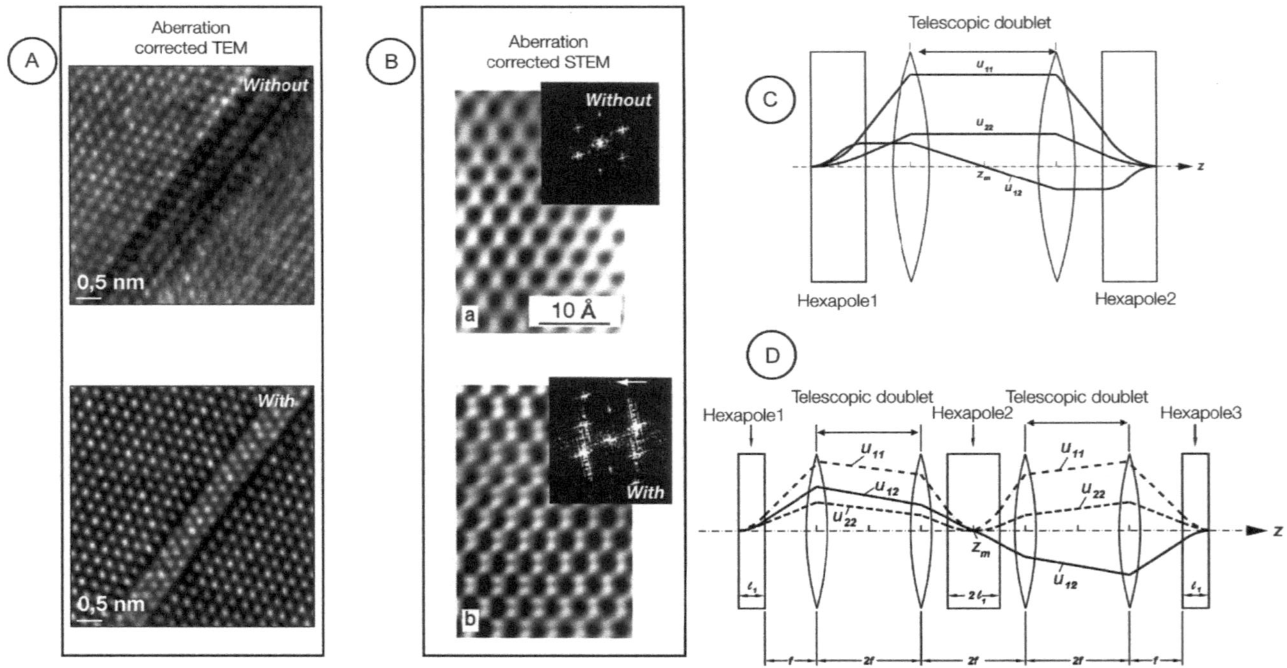

Fig. 1.3 (A) HREM images of a precipitate in a TiAl sample taken in a 200kV aberration corrected TEM without and with the C-COR activated. (B) HAADF-STEM image of a Si foil in [011] orientation obtained at 120kV. (A) Corrector's quadrupoles on, octupoles off. (B) Corrector on fully. *Insets* show diffractograms; the *arrow* marks the Si (400) reflection at ($0.136\,nm^{-1}$). The benefits of the corrector are clearly visible in both case. (C) Course of the second-order fundamental rays u_{11}, u_{12} and u_{22} within the semiaplanatic hexapole corrector (without first transfer lenses). In that case, the three terms represents the second-order deviation generated by the hexapole relatively to one paraxial trajectory, and written using a linear combination of the principal ($h(z)$) and the marginal ray ($k(z)$) following: $u(z) = C_h h(z) + C_k k(z)$, and given by $\Delta u = C_k^2 u_{11} + C_h C_k u_{12} + C_k^2 u_{22}$. (D) Course of the secondary fundamental rays within the aplanatic corrector B-COR. The two doublets of short hexapoles installed symmetrically on both sides of the intermediate image planes, located in the center of the two transfer doublets, are not represented. Only the three long hexapoles are shown. *Panel (B): Extracted from Dellby, N., Krivanek, O.L., Nellist, P.D., Batson, P.E., Lupini, A.R., 2001. Progress in aberration-corrected scanning transmission electron microscopy. Microscopy 50 (3), 177–185. Panel (D): Both layout have been extracted from Rose, H.H., 2009. In: Geometrical Charged-Particle Optics. Springer Series in Optical Sciences, vol. 142. Springer, Berlin. OCLC: ocn268931877.*

through many intermediate illumination conditions that are not really interesting to describe.

Electrons microscopes which have the ability to scan a focused probe on the sample are called Scanning TEM or STEM (Pennycook and Nellist, 2011). Since a thin sample is used, the probe spreading within the sample is relatively small, and the spatial resolution of the STEM is predominantly controlled by the size of the probe. In TEM, the optics after the sample will primarily control the properties of the microscope, whilst those before the sample are critical for creating the STEM probe. To form a small and intense probe we clearly need a correspondingly small and intense electron source such as field emission source. The physical quantity very important for an electron gun is indeed the source brightness, which will be discussed in detail in the following parts. The size of the source image is not the only probe size defining factor. As in the case of the TEM imaging objective lens, aberrations of the probe forming objective lens of STEM have greatest effect, and limit the width of the beam that may pass through the objective lens and still contribute to a small probe. Such "aberrated" beams will not be focused at the correct probe position, and will lead to large diffuse illumination thereby destroying the spatial resolution. The key benefits of spherical aberration correction in STEM are illustrated by Fig. 1.3B. When spherical aberration is corrected, it is no longer required to exclude beams that would have been significantly "aberrated," allowing for the use of a wider objective aperture. A larger objective aperture has two results:

1. the diffraction-limited probe size is smaller so the spatial resolution of the microscope is increased
2. it enables a higher current in the same probe size when the electron source size is dominating

The quadrupole–octopole corrector is used in STEM (Dellby et al., 2001; Pennycook and Nellist, 2011). Semiaplanatic C-COR hexapoles configuration can also be used in STEM to correct the probe-forming optics. However a fully aplanatic system B-COR is not necessary for STEM, where the field of view, generated by the beam scan, is not perturbed by the objective lens off-axial geometrical aberrations contrary to TEM imaging.

1.1.2 The complementarity of the wave approach

Much of the behavior of electron-optical instruments can be understood satisfactorily in terms of geometric optics as previously reported, but as soon an any wavelength-dependent phenomena need to be included, wave optics is indispensable. In addition to the electron trajectory, we have to introduce an additional concept, the geometrical wavefront.

In light optics, the concept of a geometrical wavefront appears in the work of Fermat, Malus and Hamilton as a surface of constant optical path perpendicular to all rays. This concept is used in the Abbe theory of imaging of light optics. Abbe described the optical image formation as an interference phenomenon. The image intensity is interpreted by the interference of the coherent waves, which are formed by emission or diffraction at the object, transmitted and transformed by the optical imaging system (Born and Wolf, 1999).

The Abbe theory of imaging is the same in electron imaging system and the resolution of an electron microscope should hence be limited by diffraction in aplanatic systems without chromatic aberration. Hence, for a complete description of physical image formation it is necessary to establish the transfer between geometrical ray optics and geometrical wavefronts.

1.1.2.1 Geometrical wavefronts in conventional optics

Coherent wave fields that follow the Malus–Dupin law can convert light from rays to waves and vice versa thanks to a property known as orthotomy (Gross, 2005). Then, both types of description are equivalent and interchangeable. The conditions can be summarized as:

1. Wavefronts must be real. This wavefront must be continuous and phase singularities must not exist.
2. Since wavefronts cannot be identified in the vicinity of a caustic, conversion is typically not possible.

The demonstration of the law of Malus–Dupin can be easily obtained using the Fermat principle. Fermat stipulated that light takes the path with the shortest time of travel. This can be mathematically expressed thanks to a function V defined as the sum of infinitesimal optical paths calculated using the refractive index $n(x, y, z)$ taken along the light path:

$$\delta V = \delta \int_{P_0}^{P_1} n(x, y, z) ds = 0 \tag{1.4}$$

where ds is the infinitesimal length element $ds = \sqrt{dx^2 + dy^2 + dz^2}$. The function V is characteristic of the light path between the two points P_0 and P_1 and is then called the point characteristic function (Hamilton, 1833). The point characteristic function describes the path length of an optical system by the object and image points P_0 and P_1 as parameters. In a perfect imaging optical system there can be infinitely light paths of the same length between object and conjugated image point. For the description of the light path through optical systems, the infinitesimal length element ds is impractical. With the z-coordinate as parameter instead, it follows from 1.4 that:

$$V = \int_{P_0}^{P_1} n(x, y, z).ds = \int_{z P_0}^{z P_1} n(x, y, z)\frac{ds}{dz}dz$$
$$= \int_{z P_0}^{z P_1} L(x(z), y(z), x'(z), y'(z), z)\, dz \quad (1.5)$$

where the $'$ represents the derivative relative to z coordinate.

Knowing the element: $ds = \sqrt{dx^2 + dy^2 + dz^2} = dz.\sqrt{1 + x'^2 + y'^2}$ and the Lagrangian: $L(x(z),\ y(z),\ x'(z), y'(z),\ z) = n(x,\ y,\ z).\frac{ds}{dz} = n(x, y, z).\sqrt{1 + x'^2 + y'^2}$, we can define the canonical momentum vector coordinates $\vec{p}$:

$$p_x = \frac{\partial V}{\partial x} = \frac{\partial L}{\partial x'} = n.\frac{x'}{\sqrt{1 + x'^2 + y'^2}} = n.\frac{dx}{ds} = n.s_x$$
$$p_y = \frac{\partial V}{\partial y} = \frac{\partial L}{\partial y'} = n.\frac{y'}{\sqrt{1 + x'^2 + y'^2}} = n.\frac{dy}{ds} = n.s_y \quad (1.6)$$
$$p_z = \frac{\partial V}{\partial z} = n.\frac{z'}{\sqrt{1 + x'^2 + y'^2}}$$

The total differential of the point characteristic is then given by:

$$dL = \vec{p}_1\, d\vec{r}_1 - \vec{p}_0\, d\vec{r}_0 \quad (1.7)$$

Thus, in any optical system, there is a vector field of optical direction cosines for each pair of object and image points, which is given by the gradient of the path length or the point characteristic function $V : \vec{p} = \vec{\nabla} V = n\vec{s}$. The optical path length, or point characteristic function V, takes over the role of the potential in classical mechanics. The curl of a gradient field vanishes, so that for every closed path integral (Hilbert integral) we find:

$$\vec{\nabla} \wedge \vec{p} = \vec{0} \quad (1.8)$$

Eqs. (1.6) and (1.8) express the fact that light rays run perpendicular to the planes of equal light-path lengths stated previously using the law of Malus–Dupin. The planes of equal light path lengths $V = \text{const}$ are thus equivalent to the wavefronts of the wave-optical description, as reported in Fig. 1.4A.

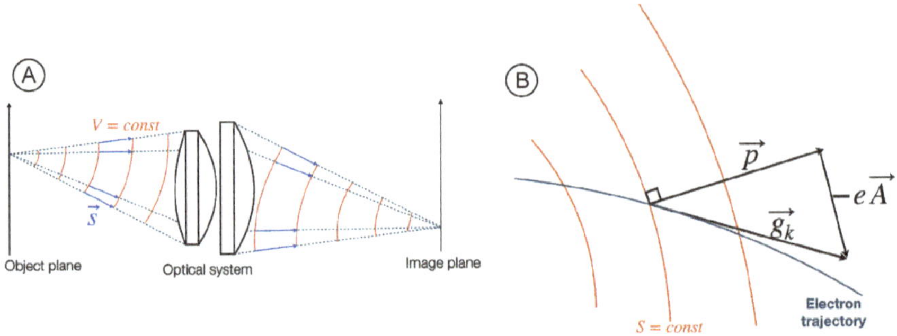

Fig. 1.4 (A) Normal congruence of rays in conventional optical system for homogeneous object and image space: the light rays run perpendicular to the wavefronts. The planes of equal V are equivalent to the wavefronts defined in wave optics. (B) Charged particle momentum (canonical and kinetic) and the surfaces of constant phase, or constant Hamilton's characteristic function S. It gives the definition of the electron wavelength in the presence of a magnetic vector potential $\vec{A}$.

1.1.2.2 Geometrical wavefronts in charged particle optics

Hamilton showed that the techniques he had developed for handling optical problems using the point characteristic function are also applicable in mechanics (Hamilton, 1833). Indeed, it is possible to define a point characteristic function for a particle moving through an electromagnetic field. This associated function is obtained most conveniently by employing the Lagrange variational principle (Rose, 2009). This principle, also known as least action principle or variational principle, requires:

$$\delta W = \delta \int_{t_0}^{t_1} L(x, y, z, \dot{x}, \dot{y}, \dot{z}, t)dt = 0$$

$$W = \int_{t_0}^{t_1} L(x, y, z, \dot{x}, \dot{y}, \dot{z}, t)dt = \int_{t_0}^{t_1} \left(\vec{p}.\vec{v} - H\left(x, y, z, p_x, p_y, p_z, t\right)\right)dt = \text{extremum} \tag{1.9}$$

where W is known as the action, L is the Lagrangian function (defined by the difference between the kinetic energy and the potential energy of the particle), H the hamiltonian function, $\vec{p} = \frac{\partial L}{\partial \vec{v}}$ defines its canonical momentum vector and $\vec{v} = \dot{\vec{r}}$ its velocity vector. The dot represents the derivative relatively to the time.

For single-particle motion through an electromagnetic field, the Lagrangian function of the electron is:

$$L = m_e c^2\left(1 - \sqrt{1 - (v^2/c^2)}\right) - e\left(\vec{v}.\vec{A} - U\right) \tag{1.10}$$

where $\vec{A}$ and U are the electromagnetic potentials, and c the speed of light.

In the case of stationary fields, which will be always considered in the following, the Lagrangian function is then stationary. Under this condition, we can easily demonstrate thanks to the Lagrangian derivative, that the corresponding Hamiltonian defines the total energy of the particle $H = E$ (sum of kinetic energy and potential energy of the particle), which is then constant. The variational principle for constant total energy is called the principle of Maupertuis and become:

$$\begin{aligned} W &= \int_{t_0}^{t_1} (\vec{p}.\vec{v} - H)dt = \text{extremum} \\ S &= W + E(t_0 - t_1) = \int_{t_0}^{t_1} \vec{p}.\vec{v}dt = \int_{t_0}^{t_1} \vec{p}d\vec{r} = \text{extremum} \end{aligned} \tag{1.11}$$

Using the Maupertuis principle, we can extract the point characteristic function S of the moving particle, by writing the equivalent of the Fermat principle for a particle. This is known as Hamilton's principle (Hawkes and Kasper, 2018; Rose, 2009; Wollnik, 2014):

$$\delta S = \delta\int_{\vec{r}_0}^{\vec{r}_1} \vec{p}d\vec{r} = \delta\int_{s_0}^{s_1} \vec{p}\frac{d\vec{r}}{ds}ds - \delta\int_{P_0}^{P_1} n(x, y, z)ds = 0 \tag{1.12}$$

where the refractive index for a charged particle moving through an electromagnetic field is $n = \vec{p}\frac{d\vec{r}}{ds}$ and can be written in terms of the electromagnetic potentials as follow:

$$\begin{aligned} n &= \frac{\partial L}{\partial \vec{v}}.\frac{d\vec{r}}{ds} = \left(m\vec{v} - e\vec{A}\right).\frac{d\vec{r}}{ds} = \left(\vec{g}_k - e\vec{A}\right).\frac{d\vec{r}}{ds} \\ &= |\vec{g}_k|\cos(\alpha) - e\vec{A}.\frac{d\vec{r}}{ds} = \sqrt{2m(E + eU)} - e\vec{A}.\frac{d\vec{r}}{ds} \end{aligned} \tag{1.13}$$

The angle $\cos(\alpha) = (\vec{g}_k/|\vec{g}_k|).(d\vec{r}/ds)$ represents the angle between the integration path and the direction of the kinetic momentum vector $\vec{g}_k$. The integration performed to obtain the point characteristic function S is always taken coincident with the flux line of $\vec{g}_k$ leading to $\cos(\alpha) = 1$.

The electromagnetic field acts as an inhomogeneous anisotropic medium of refraction for the charged particles, using an analogy with light optics (Born and Wolf, 1999; Gross, 2005). The anisotropy stems from the directional dependence of n on the direction of flight of the particle in the presence of a magnetic field. All electron lenses are gradient-index lenses according to the language of light optics because electromagnetic potentials are continuous functions of spatial coordinates and cannot abruptly change at a given surface, unlike the light-refractive index of refraction at the surface of a lens.

Going back to the wavefront, using de Broglie formulation of the electron wave, we can define the electron wave-vector $\vec{k}$ using the canonical momentum vector and the point characteristic function S:

$$\vec{k} = \frac{m\vec{v} - e\vec{A}}{\hbar} = \frac{\vec{\nabla} S}{\vec{\hbar}} \tag{1.14}$$

By assuming static electromagnetic fields, we can express the electron wavefunction through the form:

$$\Psi(\vec{r}, t) = \psi(\vec{r})\, e^{-i\omega t} = A(\vec{r})\, e^{iS/\hbar} e^{-i\omega t} = A(\vec{r})\, e^{i\int \vec{k} d\vec{r}} e^{-i\omega t} \tag{1.15}$$

where ω is the angular frequency. The electron phase reported in Eq. (1.15) cannot be measured because each component of the wavevector $\vec{k}$ depends on the magnetic vector potential $\vec{A}$. Any arbitrary component A_j can only be calculated up to the derivative because it depends on the gauge choice. As a result, using an interferometry experiment, it is only possible to extract a phase difference, rather than the electron wave's absolute phase (Hawkes and Kasper, 2018). Furthermore, if we apply the differential operator to the wavefunction defines by Eq. (1.15) we find:

$$-i\hbar\vec{\nabla}\Psi(\vec{r}, t) = \vec{p}.\Psi(\vec{r}, t) \tag{1.16}$$

Geometrically, this means that the canonical momentum vector $\vec{p}$ is perpendicular to the surfaces of constant phase. Therefore, as represented in Fig. 1.4B, in the electrostatic case, where $\vec{A} = \vec{0}$, the wavefront surfaces are perpendicular to the electron trajectories, in perfect analogy with the classical Malus–Dupin law. Things become more challenging when there is a magnetic field present since the wavefront surface is now perpendicular to the canonical momentum $\vec{p} = \vec{g}_k - e\vec{A}$ rather than the trajectories. We get to the conclusion that it is not even possible to define a wavefront surface unambiguously by taking into account the arbitrary nature of the vector potential. This unique occurrence has been seen since the earliest investigation of the electron optical refraction index, which should be viewed more like a Lagrangian function useful for computing trajectories and phase differences than a function with a specific physical meaning (Rose, 2009).

1.1.2.3 The Aharonov–Bohm effect and off-axis electron holography

1.1.2.3.1 Principle of interferometry

When two or more beams are superimposed, the distribution of intensity can no longer in general be described using a simple superimposition of the geometrical configuration of individual beams. If the superimposed beams originate in the same source, the fluctuations in the beams are in general correlated, and they are said to be completely of partially **coherent** depending on whether the correlation is complete or partial (Born and Wolf, 1999). When considering this type of illumination, the superimposition will generate interference phenomenon between each beams wave. If the beams come from difference sources, they are said to be mutually **incoherent**. Using incoherent illumination, the superimposition could be interpreted simply by the sum of the individual beam intensities without introducing interference phenomenon.

There are two general methods of obtaining many beams from single beam. The first one, equivalent to the Young experiment configuration, divides the incoming beam by the passage through apertures placed side by side. This method is called division of wavefront, and is efficient only using sufficiently small sources. In the second method, the beam is divided at one or more partially transmitted and reflected beams. This method is called division of amplitude, and can be used even with extended sources. Furthermore, we have to consider separately the effects coming from two-beam superposition and those coming from the superposition of many beams, called N-beam interferences.

Inside an electron microscope all these methods can be implemented. The diffraction of the electron beam by a crystalline sample, briefly introduced in Section 1.1.1, corresponds to a division of amplitude configuration. If only one Bragg diffracted beam direction is generated under a specific orientation between the crystal and the incoming beam, we are then in two-beam configuration. As a consequence, when the two beams are superimposed in the image plane by the action of the objective lens, we can observe an interference fringes pattern. We can mathematically calculate this pattern very simply by considering two monochromatic wave-functions

$\Psi_1(\vec{r}, t) = A_1(\vec{r}) e^{i\varphi_1} e^{-i\omega t}$ and $\Psi_2(\vec{r}, t) = A_2(\vec{r}) e^{i\varphi_2} e^{-i\omega t}$ where φ are the electron wave phases described previously. The interference pattern, coming from the superimposition of the two beams can obtained by calculating the final intensity distribution:

$$\begin{aligned} I(\vec{r}) &= |\Psi_1(\vec{r}, t) + \Psi_2(\vec{r}, t)|^2 \\ &= A_1(\vec{r})^2 + A_2(\vec{r})^2 + 2A_1A_2\cos(\phi_1 - \phi_2) \end{aligned} \tag{1.17}$$

Simple monochromatic two-beam interference exhibits intensity oscillation corresponding to an interference fringe pattern. The fringe phase corresponds to phase difference between the two electron beams, related to their difference of path length, i.e., the difference between the two characteristic functions. Going back to the division of amplitude situation generated by two-beam interferences between a transmitted and a Bragg diffracted beam, the fringe pattern observed in the image plane of an aplanatic objective lens will directly correspond to the projection of the lattice plane (*hkl*) at the origin of the diffracted beam. If the crystal is now oriented in zone axis, an N-beam interference pattern will be generated in the image plane of the objective lens. Still considering an aplanatic objective lens, the corresponding pattern will then correspond to the atomic column projection along the selected zone axis orientation. These kind of atomic resolution images of a crystalline structure observed in TEM are called high resolution images (HREM), and the contrast formation mechanism is known as phase contrast (Spence, 2013).

Division of wavefront can be also implemented in an electron microscope, but requires a small and bright electron sources. Most of the time this configuration is used to retrieve the electron phase modified during the interaction with the sample. Indeed, the phase information is lost when the intensity of a single beam is recorded, given by: $I(\vec{r}) = |\Psi(\vec{r}, t)|^2 = |A(\vec{r})|^2$ Furthermore, as already discussed, it is not physically possible to measure the absolute phase of the electron wave due to gauge choice issue.

By generating a two-beam interference between the sample beam and a reference beam, carefully chosen to have a fixed phase φ_{ref}, it becomes possible to measure their phase difference using a fringes pattern, using Eq. (1.17), and then retrieve the sample phase. This method can be easily implemented using a division of wavefront configuration known as electron holography (Tonomura, 2011).

1.1.2.3.2 Principle of electron holography

In 1948, Gabor proposed the in-line electron holography configuration as a new technique to improve the resolving power of the electron microscope which was at that time about 1.2 nm (Gabor, 1948). By superimposing the wave-function $\Psi_S(\vec{r}, t)$ propagated from the sample plane S to the recording plane, and a coherent reference wave $\Psi_{ref}(\vec{r}, t)$ of uniform intensity, an electron hologram can be recorded defined by:

$$I_{holo}(\vec{r}) = 1 + A_S(\vec{r})^2 + 2A_S(\vec{r})\cos(\varphi_S - \varphi_{ref}) \quad (1.18)$$

The in line configuration is obtained by recording the sample wave together with the reference wave, propagating along the same optic axis, in a plane located below the object plane (or any other conjugate plane) where the superimposition can be observed enabling the interference fringes to be recorded. This corresponds basically to a defocused image recorded using coherent illumination. The first attempts to carry out this experiment in the electron microscope did not fulfill the expectations, owing to the poor stability of the instrument and the low brightness of the electron gun. The breakthrough was linked to the introduction in electron microscopy of high-brightness sources, such as field-emission guns (FEGs), originally developed by Crewe and coworkers in Chicago (Crewe et al., 1968), and implemented in electron holography by the Tonomura team in Hitachi (Howie, 2012). Additionally, the development of the off-axis configuration used to record the hologram in an optical plane conjugate to the object plane, enables holography to become a very attractive method combining high spatial resolution and good sensitivity of the recorded phase (Tonomura, 2011).

Off-axis electron holograms can be obtained using an electron optical element called an electron biprism described in Fig. 1.5A (Möllenstedt and Düker, 1956).

An electron biprism is a long conducting wire of radius r charged at a voltage V_f and placed symmetrically between two plates that are kept at ground potential. In Fig. 1.5A the axis of the wire is taken in the y-direction, normal to the page, and the electrons are assumed to travel parallel to the z-axis. The two-dimensional field distribution obtained in the region around the wire can be described, in a very good approximation, to a cylindrical condenser with an outer electrode of radius r, which is of the same order of magnitude as the distance between the two plates. We can show that although electrons passing on one side of the wire are deflected at the same angle, those passing on the other side are deflected at the opposite angle (Völkl et al., 1999). Furthermore, this deflection is proportional to the potential V_f applied to the wire. It follows that for small angles, electron trajectories originally coming from a point source S, located at a distance a of the electron biprism, will seems to come from two points sources S_1 and S_2 separated by a lateral distance of $d = 2a\alpha$. This corresponds to the electron optical analogue of conventional Fresnel biprism.

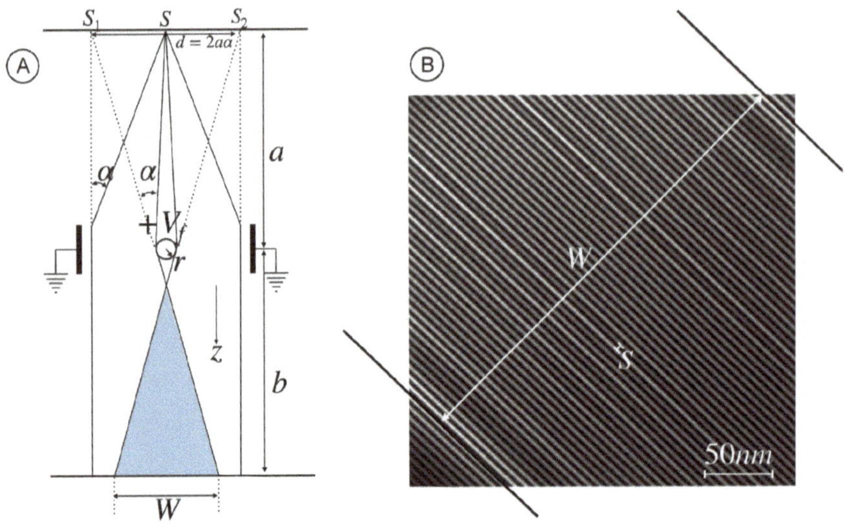

Fig. 1.5 (A) General outline describing the electron biprism analog to the well-known Fresnel biprism used in conventional optics. (B) Classical electron hologram recorded in the vacuum. *W* represents the hologram field of view and *s* the interfringes distance which will define the reconstructed phase spatial resolution.

If V_f is positive, the electron biprism will be convergent and the two wavefronts coming from S_1 and S_2 will be superimposed in an optical plane located below the wire. If V_f is negative, the electron biprism will be divergent and the two wavefronts coming from S_1 and S_2 will be virtually superimposed in an optical plane located above the wire, which could then be imaged in a real plane using an optic located after the biprism. If the source is coherent enough, interference fringes can be recorded following Eq. (1.18). Fig. 1.5B shows a typical electron hologram recorded without sample using the off-axis configuration. The figure also reports the major parameters defining the hologram: W the hologram width and s the interfringes distance, both related to V_f and the geometrical parameters of the biprism, i.e., the distances from the wire to the source a and to the recording plane b (Völkl et al., 1999).

In order to obtain amplitude and phase information, the off-axis electron hologram is first Fourier transformed. The complex Fourier transform $TF(I_{\text{holo}}(\vec{r}))$ of Eq. (1.18) can be written:

$$TF(I_{\text{holo}}(\vec{r})) = \delta(\vec{q}) + TF\left(A_S(\vec{r})^2\right) + \delta(\vec{q} + \vec{q}_c) \otimes TF\left(A_S(\vec{r})e^{i\varphi_S}\right) + \delta(\vec{q} - \vec{q}_c) \otimes TF\left(A_S(\vec{r})\,e^{-i\varphi_S}\right) \quad (1.19)$$

where the tilt of the reference wave, related to his phase φ_{ref}, is specified by the two-dimensional reciprocal space vector $\vec{q} = \vec{q}_c$. This equation describes a peak at the reciprocal space origin, corresponding to the Fourier transform of the reference image, a second peak, also centered on the origin, corresponding to the Fourier transform of a conventional TEM image of the sample, called the centerband. Then, we found a peak centered at $\vec{q} = \vec{q}_c$ corresponding to the Fourier transform of the wave-function $\Psi_S(\vec{r}, t)$, and a second peak centered at $\vec{q} = -\vec{q}_c$ corresponding to the Fourier transform of the complex conjugate of the wave-function $\Psi_S(\vec{r}, t)$, called sidebands. In order to recover the amplitude and the relative phase shift φ_S of the electron wave-function, one of the two sidebands is digitally selected and inverse Fourier transformed. Electrostatic and magnetostatic configuration of the sample can be extracted from the electron phase, following the Aharonov–Bohm effect (Hawkes and Kasper, 2018; Tonomura, 2011; Völkl et al., 1999).

1.1.2.3.3 The Aharonov–Bohm effect

As described previously, the refractive index in electron optics contains the vector potential and not the magnetic field strength. Therefore, as commented by Ehrenberg and Siday in 1949 "one might expect wave-optical phenomena to arise which are due to the presence of a magnetic field but not due to the magnetic field itself, i.e., which arise while the rays are in field-free regions only" (Ehrenberg and Siday, 1949). As a result, they suggested doing a two-beam interference experiment in a wavefront division configuration with two beams passing on either side of an enclosed flux, and they assumed that the flux would produce a detectable phase shift. The Ehrenberg–Siday configuration could be implemented in a TEM using the off axis electron holography experiment, where the biprism could for instance split electrons trajectories 1 and 2 before the enclosed flux (Hawkes and Kasper, 2018). Considering the electron optical refraction index reported in Eq. (1.13), the phase difference between the two trajectories 1 and 2 can be written:

$$\begin{aligned}\Delta\varphi &= \frac{1}{\hbar}\int_1 nds - \frac{1}{\hbar}\int_2 nds = -\frac{e}{\hbar}\int_1 \vec{A}d\vec{r} + \frac{e}{\hbar}\int_2 \vec{A}d\vec{r} \\ &= -\frac{e}{\hbar}\oint \vec{A}d\vec{r} = -\frac{e}{\hbar}\Phi \end{aligned} \qquad (1.20)$$

This phase variation acts on the whole interference by a lateral displacement Δx of the fringes induced by the magnetic flux Φ, with respect to the position where $\Phi = 0$. This result is gauge independent because the flux Φ is related to the circulation of the vector potential. This phenomenon then corresponds to the real physical effect which can be measured in the electron phase shift (Tonomura et al., 1986).

Aharonov and Bohm suggested a second electron interference experiment to examine the impact of the electric potential on the electron phase (Aharonov and Bohm, 1959). Such as for the magnetic case, a coherent electron beam is split into two parts, using for instance a first electron biprism. The two beams will have to go through two empty metallic cylinders of length l. Only when the two wavepackets are located inside the cylinders, different voltage values are applied to the cylinders, so that the electron do not experience the action of the electric field. The beams are finally recombined using a second biprism allowing to generate interference fringes. The phase shift induced by the voltage difference between the two tubes can be expressed as follow:

$$\Delta\varphi = \frac{\pi l \Delta V}{\lambda V_0} \tag{1.21}$$

where V_0, in the nonrelativistic case, is the accelerating potential. Using a continuous electron beam, the electric Aharonov–Bohm effect could be observed applying a time dependent voltage to the two tubes. However, this required a highly sophisticated setup, and as far as I know, nobody succeeded to implement this idea inside a TEM. Maybe the use of ultrafast coherent electron beams, recently developed in the laboratory, will open a new direction to observe such fundamental effect. These recent works will be discussed in the part dedicated to the development of ultrafast coherent source.

1.1.2.3.4 Consequence for material sciences

Off axis electron holography can be implemented to retrieve the phase shift between the sample and the reference waves (Völkl et al., 1999). Following the Aharonov–Bohm effect, the measurement contains information about the sample electrostatic potential (such as the mean inner potential related to the composition, density and/or presence of an excess of electric charges) and the magnetic vector potential if the sample is magnetic. Therefore, it becomes possible to quantitatively map the electric and the magnetic properties of materials at the nanoscale using electron holography set in off-axis configuration inside a TEM, if the electrons beam is coherent enough. It is important to stress that the total phase shift will finally correspond to an integration of the electromagnetic potentials along the electron path.

1.1.3 Key instrumental figures in TEM

1.1.3.1 Influence of geometrical aberrations of lenses

We already met the general concept of aberrations in an optical system as the failure of the system to conform to the mode of ideal image formation described using paraxial properties (Hawkes and Kasper, 2018). There are several methods used to describe how the rays miss the image point. Very often it is convenient to think in terms of transverse aberrations (Gross, 2005). First, let's consider a single object point and a given position of the image plane and on this plane consider a given reference point, which is the point where the image should be. It is worth noting that the position of the image plane as well as the choice of the image reference point may differ from the paraxial values if required. Thanks to the power expansion of the trajectory described previously in Eq. (1.3), we can identify usual transverse aberrations as nonlinear coefficients (.|.) of the expansion (1.3) taken at the position z of the image plane (Wollnik, 2014).

Transverse aberrations described in the image plane, can be also represented in terms of wavefront aberrations in the exit pupil plane which is more convenient, especially dealing with interferometric methods (Gross, 2005). As reported in Fig. 1.4A in object space, where all rays pass through the object point, the wavefronts have the shape of concentric spheres, with the object point as center. In image space, when there are no aberrations at all, the wavefronts are again spheres, centered on the image point. In the case of aberrations, the wavefronts in the image space are no longer spheres, and the deviation to a suitable reference sphere is the wave aberration. As described in Fig. 1.6A, the radius R of the reference sphere is chosen so that it contains the intersection point of the principal ray with the optic axis, i.e., the point Q defining the location of the exit pupil.

Using Hamilton's characteristic function S_{OB} defining the equivalent of the optical length from the object point O to a point B located in the imaging space, we know that:

$$S_{OB} = S_{OQ} \tag{1.22}$$

as B and Q are points on the same wavefront. So, the wavefront aberration W is defined as the optical path along the ray from the reference sphere to the wavefront (Welford, 1989):

$$W = S_{OB} - S_{OP} = S_{OQ} - S_{OP} \tag{1.23}$$

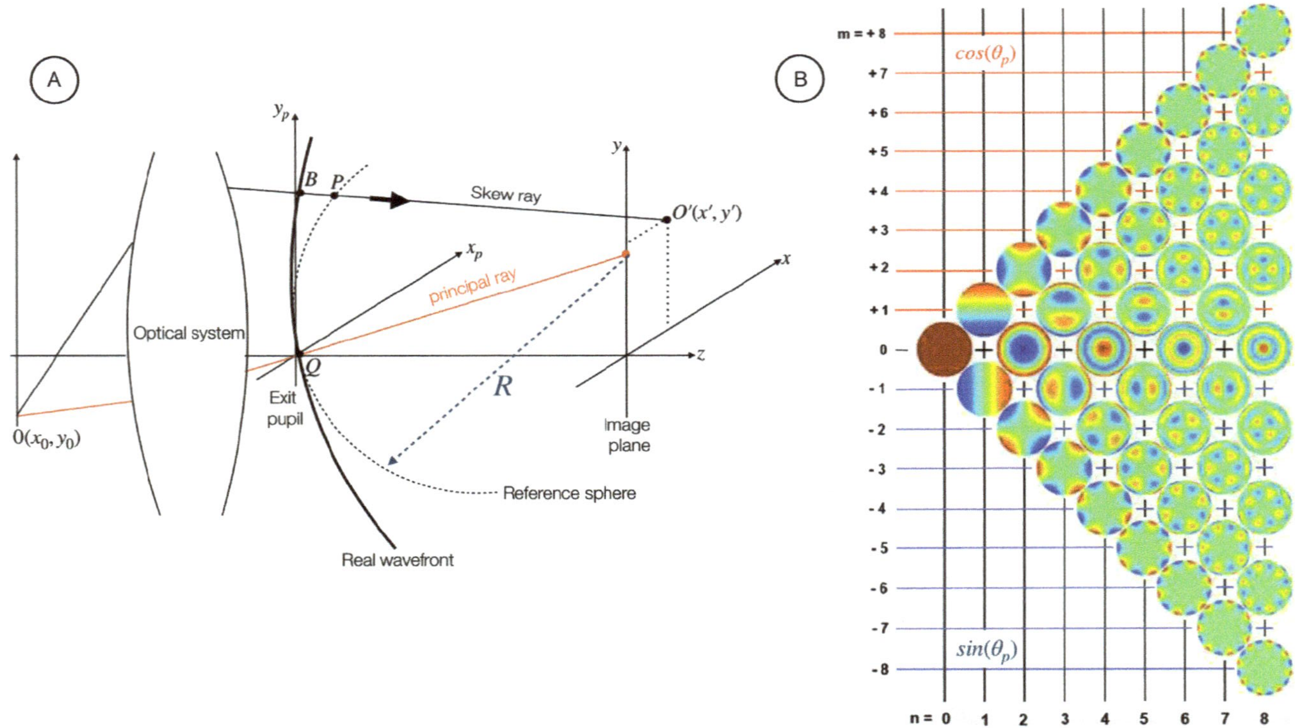

Fig. 1.6 (A) Description of optical aberrations as wavefront aberrations $W = [BP]$ measured in the exit pupil plane of the optical system. (B) Pictures representing all the coefficients of the Zernike polynomials used to fit the wavefront aberrations determined in the exit pupil plane. *From Gross, H. (Ed.), 2005. Handbook of Optical Systems. Wiley-VCH, Weinheim. OCLC: ocm56963682.*

When the wavefront aberration for all rays emerging from the object point O and passing through the exit pupil is to be described, the rays concerned are identified by their pupil coordinates x_p and y_p and the wavefront aberration $W(x_p, y_p)$ is a function of two variables. The pupil coordinates can be expressed using the angular components a and b previously introduced; they are strictly equivalent. It must be pointed out that this function $W(x_p, y_p)$ only describes the aberrational behavior for the fixed chosen object point. If we characterize an object point by its object plane coordinates x_0, y_0, then the wave aberration becomes a function of four variables $W(x_0, y_0, x_p, y_p)$. In the same way as for the trajectory coordinates $x(z)$, $y(z)$, …, the geometrical wavefront aberration function can now be expanded as a power series in each variable, arranging the terms in the proper order following;

$$W(x, x_p, y, y_p) \sum_{k,l,m,n} a_{klmn} x_p^k y_p^l x_0^m y_0^n \quad (1.24)$$

This power expansion is strictly mathematical. As the wavefront aberration is not an arbitrary function to be expanded, but is defined in such a way that it vanishes at the pupil center ($x_p = y_p = 0$), all coefficients a_{klmn} in expressions with no dependence on the pupil coordinates must be zero. It is well known that the transverse ray geometrical aberrations coefficients $(x|aaa)$, $(x|aax)$, … can be calculated to a good approximation by differentiating the wavefront aberration with respect to x_p and y_p (Gross, 2005).

Furthermore, in the case of circular aperture, it is convenient to describe the wavefront aberrations using Zernike polynomials. The Zernike polynomials are defined in the exit pupil. The Zernike polynomials are basically defined in polar coordinates and consist of a radial term $R(r_p)$ and a term dependent on the azimuthal angle θ_p (Gross, 2005):

$$W(r_p, \theta_p) = \sum_{n} \sum_{m=-n}^{n} c_{nm} Z_n^m(r_p, \theta_p) \quad (1.25)$$

with

$$\begin{aligned} Z_n^m(r_p, \theta_p) &= R_n^m(r_p).\sin(m\theta_p) \text{ with } m < 0 \\ Z_n^m(r_p, \theta_p) &= R_n^m(r_p).\cos(m\theta_p) \text{ with } m > 0 \\ Z_n^m(r_p, \theta_p) &= R_n^m(r_p).1 \text{ with } m = 0 \end{aligned} \quad (1.26)$$

and the radial function defined by $R_n^m(r_p) = \sum_{k=0}^{(n-m)/2} \frac{(-1)^k (n-k)!}{k!((n+m)/2-k)\ !((n-m)/2-k)!} r_p^{n-2k}$.

They represent a complete set of surface deformations by which an arbitrary wavefront aberration can be expanded into discrete shapes of definite size. This makes it possible to classify more easily the wavefront aberrations power expansions and to quantitatively describe the surface deformations. As reported in Fig. 1.6B they are classified using a radial index n and an azimuthal one m (Gross, 2005).

Following Eq. (1.15), the wavefront aberration function will induce a phase shift of the electron wave described by the coherent transfer function (WTF):

$$\mathrm{WTF}\left(r_p, \theta_p\right) = B\left(r_p, \theta_p\right) e^{-iW\left(r_p,\theta_p\right)/\hbar} \tag{1.27}$$

where $B(r_p, \theta_p)$ represents the aperture stop function in the pupil plane sometimes called pupil function. From here onward, the influence of such a function on the image processes will be separated between TEM and STEM imaging. Fig. 1.2 reports the general setup of the two descriptions.

1.1.3.1.1 TEM imaging in the presence of aberrations

The Abbe theory of imaging used a Fourier description of the optical image formation as described in Fig. 1.7 (Gross, 2005).

Starting from the effective light source $s(\nu_s)$, a parallel illumination wave is obtained in the object plane using the Fourier transform of the light source $S(x) = TF(s(\nu_s))$. The object can be described in approximation by a complex transfer function $T(x)$. The object wave encountered directly behind the object will be $U_0(x) = T(x).S(x)$. In the objective lens pupil plane,

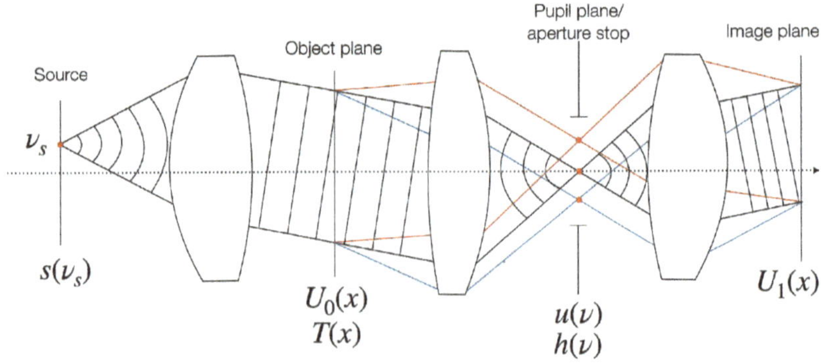

Fig. 1.7 Abbe theory of imaging: interpretation of optical imaging, generated by a 6f-optical system, using diffraction and interference. Mathematically the image formation is easily described using Fourier sequences between image and pupil planes.

described using the generalized coordinate ν, there is then the Fourier transform of the object wave $u_o(\nu) = TF(U_0(x)) = t(\nu) \otimes s(\nu)$. Now, in this pupil plane the filtering effect of the aperture stop and the wavefront geometrical aberrations, described by Eq. (1.27), takes place. Thus the wave distribution behind the aperture, and in the image plane will be given respectively by:

$$u_1(\nu) = \mathrm{WTF}(\nu).u_o(\nu) = \mathrm{WTF}(\nu).t(\nu) \otimes s(\nu)$$
$$U_1(x) = \mathrm{FT}(\mathrm{WTF}(\nu)) \otimes [T(x).S(x)] = \mathrm{FT}(\mathrm{WTF}(\nu)) \otimes [U_0(x)] \quad (1.28)$$

The intensity distribution of the final image is given by $I_1(x) = |U_1(x)|^2 = |\mathrm{FT}(\mathrm{WTF}(\nu)) \otimes [U_0(x)]|^2$. Then due to aberrations, there is no linear correlation between the image intensity I_1 and the object intensity $I_0(x) = |U_0(x)|^2$. That's the reason why in HREM imaging, aberrations have to be considered for the interpretation of the atomic contrast (Spence, 2013).

1.1.3.1.2 STEM imaging in the presence of aberrations

As reported in Fig. 1.2B, the illumination system of a STEM is set to form a reduced image of the electron source on the sample plane. The smallest probes can only be obtained using fully coherent illumination. The coherent length L is measured in the aperture stop plane, defined by one pupil plane of the illumination system. A fully coherent illumination is obtained when L is bigger than the real aperture size R_a.

Considering an "incoherent" conditions (i.e., $L \ll R_a$), the total probe diameter $d_{total}^2 = d_s^2 + d_d^2 + d_{GA}^2 + d_c^2$ can be easily calculated by adding in quadrature all the various contributions to the paraxial image of the electron probe d_s: transverse geometrical aberrations d_{GA}, axial chromatic aberrations d_c and the diffraction limited probe d_d (Spence and Zuo, 2013). The smallest probe is then simply obtained by minimizing these quantities.

The Abbe theory of imaging described previously and applied to self-luminous object, can be used to understand the STEM probe formation using coherent illumination (Hawkes and Spence, 2007). In that case we must consider the electron source wave function, described by $S(x)$, in the same way as the object wave in the description of TEM imaging. As a consequence, the actual image of the probe in the sample plane will be defined by a convolution of $S(x)$ and $\mathrm{FT}(\mathrm{WTF}(\nu))$. The final spot intensity is found by taking $I_1(x) = |S_1(x)|^2 = |\mathrm{FT}(\mathrm{WTF}(\nu)) \otimes [S(x)]|^2$. In conventional STEM instrument, the probe is mainly limited by the spherical aberration (Pennycook and Nellist, 2011). In the case of an aberration-corrected instrument, the limiting aberration is unlikely to be spherical aberration. Uncorrected higher-order aberrations, or indeed lower-order aberrations produced parasitically, may be limiting. Unsurprisingly, aberration correction is capable of producing much smaller probes than an uncorrected instrument, but in both case, like for TEM imaging, the influence of aberrations in the formation of the probe have to be considered to understand the image formation mechanism (Pennycook and Nellist, 2011).

1.1.3.2 The source and the electron coherence

In order to further investigate the influence of the electron source on the image intensity, a single delta-like light source point p at coordinate ν_p is considered, and $S(x)$ in Eq. (1.28) is given by a plane wave. If we note $H(x) = \mathrm{FT}(\mathrm{WTF}(\nu))$, we can define the normalized image amplitude U_p for a source point p:

$$U_p(x, \nu_p) = \mathrm{FT}(\mathrm{WTF}(\nu)) \otimes \left[T(x).e^{2\pi\nu_p x}\right] = \int T(x').H(x - x')\, e^{2\pi\nu_p x'}\, dx' \tag{1.29}$$

The image intensity is given by the squared value of the integral over all image amplitudes to all source points:

$$I(x) = \left| \int s(\nu_p).U_p(x, \nu_p) d\nu_p \right|^2 \tag{1.30}$$

After insertion of Eq. (1.29) one formally gets a quadruple integral for the image intensity consisting of a twofold convolution and the double integral over all source points (Gross, 2005):

$$I(x) = \int s(\nu_p). \int T(x_1).H(x - x_1)e^{2\pi\nu_p x_1} dx_1\, d\nu_p. \int s^*(\nu_q). \int T^*(x_2).H^*(x - x_2)e^{-2\pi\nu_q x_2} dx_2\, d\nu_q \tag{1.31}$$

where the star stands for the complex conjugate. In the double source integral in Eq. (1.31) the mutual coherence function of the source Γ_s at the position of the object can be found, which is also called the mutual intensity of the illumination distribution:

$$\Gamma_s(x_1, x_2) = \iint s(\nu_p)\, s^*(\nu_q)\, e^{2\pi(\nu_p x_1 - \nu_q x_2)}\, d\nu_p d\nu_q \tag{1.32}$$

Eq. (1.30) could then be rewritten using the mutual coherence function:

$$I(x) = \int\int \Gamma_s(x_1, x_2)\, T(x_1).H(x - x_1)\, T^*(x_2).H^*(x - x_2)dx_1\, dx_2 \tag{1.33}$$

Considering an uncorrelated electron source, interference contributions of different source points vanish due to the averaging in time of the intensity formation. Only those interference terms that come from one and the same source point, and thus have a rigid phase correlation to each other, will remain. Under this assumption it follows for Eq. (1.32) that the mutual coherence function is given by the Fourier transform of the intensity distribution of the effective source (Zernike, 1938):

$$\begin{aligned}\Gamma_s(x_1, x_2) &= \int\int s(\nu_p)\, s^*(\nu_q)\, \delta(\nu_p - \nu_q)\, e^{2\pi(\nu_p x_1 - \nu_q x_2)}\, d\nu_p d\nu_q \\ &= \int |s(\nu_q)|^2\, e^{2\pi\nu_q(x_1 - x_2)}\, d\nu_q\end{aligned} \tag{1.34}$$

Depending on the extension of the source, the imaging is called incoherent, coherent or partially coherent. As discussed previously for the STEM case, considering a large electron source size will generate small coherence length L in the pupil plane. If L is much smaller than the diameter of the aperture, it is always an incoherent imaging process. On the other hand, for a point-like electron source, the object is illuminated only by a plane wave. It is then always called a coherent image. For electron source dimensions in between, it is called a partially coherent image (Gross, 2005).

Let's now investigate the influence of the mutual coherence function Γ_s in the formation of off-axis electron holograms using partially coherent illumination. We first introduce the self-coherence of the electron beam at x_1. It reduces to the ordinary intensity of the beam taken at this point $\Gamma_1(x_1, x_1) = I_1 = A_1^2$. The complex degree of coherence of the beam is defined by the normalization of mutual coherence function $\Gamma_s(x_1, x_2)$ (Lichte, 1991).

$$\gamma_s(x_1, x_2) = \Gamma_s(x_1, x_2)/(A_1 \times A_2) = \Gamma_s(x_1, x_2)/\left(\sqrt{\Gamma_1} \times \sqrt{\Gamma_2}\right) \tag{1.35}$$

$\gamma_s(x_1, x_2)$ represents the *spatial* complex degree of coherence of the beam between two points x_1 and x_2 located after the source. The same argumentation can be implemented considering the *temporal* complex degree of coherence $\gamma_t(t, t+\tau)$. This quantity can be defined using the temporal mutual coherence function $\Gamma_t = \mathrm{FT}(s(k))$ given by the Fourier transform of the source spectrum of different wave numbers $s(k)$ around the nominal wave number k_0 corresponding to the accelerating voltage V_0 of the electron microscope. Assuming that each source point emits the same spectrum, the complex degree of coherence is defined by $\gamma_{\text{total}} = \gamma_s.\gamma_t$. The interference pattern recorded using a partially coherent source, defined by it's complex degree of coherence, is then obtained by introducing $\gamma_{\text{total}} = |\gamma_{\text{total}}|^{i\varphi_{\text{source}}}$ inside Eq. (1.17):

$$I(\vec{r}) = A_1(\vec{r})^2 + A_2(\vec{r})^2 + 2A_1A_2|\gamma_{\text{total}}|\cos(\varphi_1 - \varphi_2 + \varphi_{\text{source}}) \quad (1.36)$$

Because of the narrow energy spread of electrons, γ_t is close to 1 for phase differences up to $2\pi \times 10^4$ (Lichte, 1991). Consequently, there is nearly no limitation for interferometry, or holography, due to temporal coherence. However, spatial coherence induces severe limitations. For a Gaussian distribution of $s(\nu_q)$, the *coherent current* available at a given γ_s is given by:

$$I_{\text{coh}} = -\ln(\gamma_s)\frac{B}{k_0^2} \quad (1.37)$$

where B is the *brightness* of the source. This number is then a fundamental property of the electron emitter. The total coherent current corresponds to the amount electrons current within the coherently illuminated cone (Lichte, 1991).

The main challenges in coherent TEM, especially in electron interferometry and holography, will be to optimize this coherent current and make an optimum use of it. The need for high-brightness sources, such as field-emission guns, is then obvious for such techniques. The influence of sample illumination conditions will also have to be carefully considered.

Usually we introduce the concept of reduced brightness which is independent of acceleration voltage (Hawkes, 2009). A practical definition of the reduced brightness requires a proper definition of the virtual source size d_v as seen from an electrode at voltage V_1 located after the electrons emitter. If the electrons beam current density is uniform over the acceptance aperture, B_r is defined by the following relationship:

$$B_r = \frac{4I'}{\pi d_v V_1} \quad (1.38)$$

where I' is the angular current density of the source which depends on the gun technology. This property will be discussed more deeply in Chapter 2 dedicated to new developments in cold FE technology.

1.1.3.3 Additional aberrations and parasitic effects

1.1.3.3.1 Chromatic aberration

We have so far concentrated on the influence of monochromatic geometrical aberrations on the imaging process in TEM and STEM, and particularly primary aberrations such as spherical aberration, coma and distortion. Chromatic aberrations correspond to all the nonlinear components containing the energy contribution δ in the coefficients $(.|.)$ of the aberration expansion reported in Eq. (1.3) (Wollnik, 2014). These effects simply comes from variation of electron lens strength depending on the electrons beam incoming energy. Indeed, all electrons sources have their intrinsic energy spread ΔE which will be at the origin of all these chromatic effects.

In our case, we will only discuss second-order contributions $(x|a\delta)$, $(x|x\delta)$. Linear chromatic terms $(x|\delta)$ are null in system containing two planes of symmetry such as round lenses or quadrupoles, etc. These two chromatic aberrations are called axial chromatic aberration $(x|a\delta)$ and chromatic distortion $(x|x\delta)$. Axial chromatic aberration affects the resolution while chromatic distortion represents the chromaticity of the magnification. Indeed, in the case of axial chromatic aberration the beam started from the axial position, and described by a side energy $\delta_0 = \Delta E/V_0$, will cross the paraxial image plane at a position $x_{im} = (x|a\delta)a_0\delta_0$. This aberration is linear in δ_0, that's why it is sometimes called first-order chromatic aberration (Rose, 2009). Like for spherical aberration, a minimal particle beam width (for a given angular spread) is formed at some least confusion plane shifted with respect to the paraxial image plane. In round lenses because of the cylindrical symmetry all axial chromatic terms $(x|a\delta)$, $(y|b\delta)$ are the same and are usually noted C_c. Hence, the effect of the axial chromatic aberration C_c is to convert an axial ideal point source into one with an apparent radius of (Hawkes, 2009):

$$\Delta r_c = C_c a_0 \frac{\Delta E}{V_0} \tag{1.39}$$

In the case of an off-axis position of the object point, the chromatic distortion $(x|x\delta)x_0\delta_0$ causes a lateral shift of the image point in the x-direction, depending on the particle energy. In the presence of a continuous energy spread in the beam this image is therefore blurred. In the case of round magnetic lenses, due to the Larmor rotation of the electrons we have to consider complex chromatic distortion coefficients defined by a complex mixing between $(x|x\delta)$ and $(y|y\delta)$. In that case, the real part of the complex chromatic distortion coefficient, usually noted D_c, accounts for the isotropic or radial chromatic distortion and the imaginary part accounts for the anisotropic or azimuthal chromatic distortion (Rose, 2009).

The influence of axial chromatic aberration in incoherent STEM imaging has been briefly discussed previously, and is taken into account by simply adding in quadrature the contributions d_c to the paraxial image given by the theoretical diameter of the spot measured at the paraxial image plane due only to chromatic aberration and calculated using Eq. (1.39) and the linear lens magnification.

In coherent imaging (TEM or STEM), the effect of axial chromatic aberrations can be incorporated by multiplying the coherent transfer function (WTF) described by Eq. (1.27) with an effective aperture function $f_c(r_p)$ defined in the exit pupil plane and given by:

$$f_c(r_p) = \exp\left(-\frac{1}{2}\left(\pi C_c \lambda_0 \frac{\Delta E}{V_0} r_p^2\right)\right) \tag{1.40}$$

The envelope function f_c is specific to the axial chromatic aberration and is only defined for one illumination angle. To take into account the illumination aperture a_0, this aperture function is multiplied by a second envelope $f_a(r_p)$. These two envelopes, sometimes called incoherent envelope function can be included in the general pupil function $B(r_p, \theta_p)$ already described in Eq. (1.27) (Hawkes and Spence, 2007).

1.1.3.3.2 Parasitic aberrations

Unlike the Seidel description of intrinsic aberrations of lenses, parasitic aberrations are defects resulting from lens imperfections or misalignment of the instrument (Hawkes and Kasper, 2018). For instance, due to imperfection in the rotational symmetry of the pole piece bore used in a round magnetic lens, first-order astigmatism will arise. First-order astigmatism should not be confused with high-order astigmatism encountered in the aberrations expansions, described in Eq. (1.3), through mixed geometrical coefficients such as third-order terms $(x|ayy)$, $(x|bxy)$, … intrinsic to any "perfect" lens. Most of the time, first-order astigmatism will be the dominant parasitic effect. Fortunately it can be easily corrected thanks to weak quadrupoles added to the round lens called a "stigmator." A rigorous description of parasitic aberrations in the framework of wavefront aberrations expansion has been realized by Uhlemann and Haider (1998).

1.1.3.3.3 Instrument instabilities

In addition to the influence of chromatic aberration originating from the energy spread of the source ΔE, temporal incoherence can also arise from variation in the lens currents, and accelerating voltage. All these effects cause the focus of the lenses to fluctuate. The final image is then the integration of the images given for each focus value. In coherent TEM imaging, these effects can be simply described in the chromatic envelope expressed using Eq. (1.39) by replacing $C_c \frac{\Delta E}{V_0}$ by $C_c \sqrt{\left(\frac{\Delta E}{V_0}\right)^2 + \left(\frac{\Delta V}{V_0}\right)^2 + 4\left(\frac{\Delta I}{I}\right)^2}$ where ΔV is the fluctuation in the incident voltage, and $\Delta I/I$ is the relative fluctuation of the lens current (Spence, 2013).

In electron holography, instrument instabilities will have drastic consequences in the signal to noise (S/N) of the reconstructed phase. Indeed, as previously described, electron holography is used to quantitatively retrieve the electron phase $\Delta\varphi$ modified during the interaction with the sample due to the Aharonov–Bohm effect. To maximize the accuracy of this technique, the phase resolution, for a given spatial resolution, must be optimized by determining the set of experimental conditions that enhance the hologram figure of merit (Harscher and Lichte, 1996). These conditions are determined by a number of competing parameters such as the source beam spatial coherence γ_{total} already discussed but also the camera modulation transfer function (MTF) and, of course, the instrument instabilities (mostly electrical and mechanical). The hologram figure of merit is given by the phase detection limit, which corresponds to the smallest detectable phase difference between adjacent pixels of the camera at a desired signal to-noise ratio $(snr)_{\varphi}$. This value can be estimated using the standard deviation of the measured phase. As already discussed by Harscher and Lichte (1996), the noise contribution in the hologram comes from the shot noise of the electron beam, described by a Poisson distribution, and the noise of the detection system. De Ruijter and Weiss showed that the standard deviation of a reconstructed phase can then be written (de Ruijter and Weiss, 1993):

$$\sigma_{\varphi} = \sqrt{\frac{2c}{C^2 N}} \tag{1.41}$$

where N is the number of electrons per pixel during the measurement time, and C is the hologram fringes contrast. C corresponds to the degree of coherence γ_{total}, introduced in Eq. (1.36), if we only take care of the source partial coherence. The constant $c = 1$ when part of the incoming electron beam propagates in vacuum. The total number of electrons per pixel N is strongly influenced by the detector quantum efficiency (DQE). This property of the detection system, is defined by $\mathrm{DQE} = \mathrm{SNR}_{\mathrm{out}}(\nu)/\mathrm{SNR}_{\mathrm{in}}(\nu)$, where $\mathrm{SNR}_{\mathrm{out}}(\nu)$ and $\mathrm{SNR}_{\mathrm{in}}(\nu)$ are signal-to-noise ratios given at input and output of the detection system measured at a spatial frequency ν. Therefore the phase detection limit as a function of the $(snr)_{\varphi}$ and the standard deviation σ_{φ} can be expressed as (Lehmann, 2004):

$$\delta\varphi = \sigma_{\varphi} \times (snr)_{\varphi} = \frac{(snr)_{\varphi}}{C}\sqrt{\frac{2}{\mathrm{DQE} \times N}} \tag{1.42}$$

where the desired signal-to-noise ratio $(snr)_{\varphi}$ is a given parameter. To properly determine the result of an experiment without ambiguity, it is generally admitted that $(snr)_{\varphi}$ should be chosen between [3–10]. To improve the phase detection limit, one needs to optimize the contrast C of the hologram together with the total number of electrons N. However, as already demonstrated, the best obtainable contrast is first given by the limited degree of coherence of the electrons beam γ_{total} linked to the beam brightness. It is worth noting that each linear reduction of contrast C has to be counterbalanced by a quadratically increasing of the total number of electrons per pixels N. Once more, this shows the central role of the beam brightness in electron holography (Tonomura, 2011).

Nevertheless, considering a given beam coherence, $\delta\varphi$ could still be optimized by increasing N using higher exposure time. However, increasing the acquisition time will automatically induce higher contribution of experimental set-up instabilities. They will then strongly decrease hologram contrast C. So even with high brightness source we have to take into account these parameters.

Lehmann et al. proposed a simple mathematical description of the influence of the beam partial coherence, the experimental instabilities, the inelastic interactions of the electron beam inside the specimen and the detector MTF, for a given interfringe distance s, on the contrast C (Lehmann, 2004):

$$C = \gamma_{\mathrm{total}} \times C_{\mathrm{inst}} \times C_{\mathrm{inel}} \times \mathrm{MTF} \tag{1.43}$$

where C_{inst} and C_{inel} correspond to the influence of experimental instabilities and inelastic scattering respectively.

1.1.3.3.4 Biprism artefacts specific to electron holography

The first well-known artefacts encountered in electron holography are the Fresnel fringes generated by the scattered electrons on the edge of the

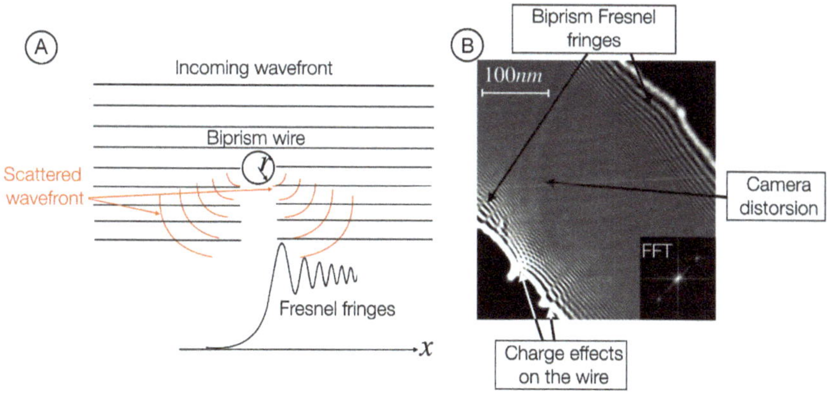

Fig. 1.8 (A) Outline showing the origin of Fresnel fringes observed in electron holography. (B) Typical artefacts encountered in electron holography.

biprism wire (Völkl et al., 1999). As described in Fig. 1.8A, these scattered wavefronts (drawn in red) will interfere with the unscattered one (in black). The resulting interference pattern, known as Fresnel fringes pattern, can be observed in an optical plane located below the wire, or above if a lens is used to focus the virtual image. As reported in Fig. 1.8B, these large Fresnel fringes are finally superimposed to the hologram fine fringes. The hologram fringes amplitude will then be affected by the Fresnel fringes. Fresnel inter-fringes distance vary continuously from the edge of the hologram to the center. They can then be easily identified in the fast Fourier transform (FFT) of the hologram by the white line joining the centerband to the two sidebands (Hawkes and Spence, 2007).

A second artefact appears due to charge effect on the biprism wire. Indeed, most of the time the wire is produced by coating an ultrasmall ($r \sim 200$ nm) quartz fiber with noble metal such as gold. Over time, degradation can be observed in the coating surface generating areas where the glass fiber could be in contact with the incoming electrons beam. In these specific zones, charges phenomenon will appear and induce strong distortion and parasitic Fresnel fringes in the final hologram as reported in Fig. 1.8B (Völkl et al., 1999).

Finally, parasitic distortions can be also generated by the detection system. In the early days, electron holograms were recorded on photographic film, but digital acquisition using charge-coupled devices (CCD) or CMOS based cameras and direct electron detectors (DED) is now widely used due to their linear response, good dynamic range and high DQE, which is a key parameter to optimize the phase detection limit, as already discussed previously. Furthermore using CCD, CMOS or DE detectors, immediate accessibility to the recorded information is also an important attractive factor. Recording an electron image using CCD or CMOS based camera first starts with conversion between electrons and photons thanks to a scintillator. The photons are then transferred to the CCD or the CMOS photon detector using an optical system which can be different and specific for each manufacturer (Williams and Barry Carter, 2008). In CEMES, we mainly use Gatan detectors. Photon transfer in Gatan devices is based on a compact assembly of optics fibers allowing to reduce the size of the camera. However such fibers assembly cannot be perfectly periodic due to the high number of pixels involving the use of billions of optics fibers (such as in $4k \times 4k$ detectors) (Denneulin et al., 2014). Error in the periodic assembly of these fibers will then be detected as an artefact in the extracted hologram phase. Other artefacts will be generated using these detectors due to error in the dark image, or the detector gain, ... as observed in the hologram of Fig. 1.8B.

All these artefacts have to be taken into account during the hologram analysis and corrected. Hence, these effects can be removed to some extent by using a reference hologram, and by Fourier-filtering the sideband before reconstruction of the image wave (Völkl et al., 1999). As we will see in the following, removal of Fresnel fringes from electron holograms is also possible using multiple-biprisms electron holography (Harada et al., 2004).

1.1.4 Summary and discussion

The general guidelines of my research work is to implement original characterization methods of materials by making an optimum use of field emission TEM. Indeed, due to the high brightness of their source, this kind of instruments are the only one which could generate enough coherent current (Hawkes, 2009).

Using coherent electron sources and the two sample illumination configurations provided by a TEM and described in Fig. 1.2, it is possible to implement either highly localized analysis such as convergent beam electron diffraction (CBED), or coherent imaging analysis such as HREM or off-axis electron holography. I was mainly interested in implementing these methods to determine structural properties of a crystal over large micron or submicron areas, to the atomic scale. These properties could be localized such as structural defects (for instance dislocations), or extended like elastic crystal strain state.

Improving the amount of coherent current will be benefits for all these methods. Indeed, as already discussed previously, higher spatial coherence will allowed to focus smaller spots in the case of CBED and will improve interference fringes contrast in electron holography. In this particular case, if the fringes contrast can be improve without modifying the exposure time, it will then automatically improve the signal over noise ratio of the reconstructed phase.

There are mainly two possible directions to increase the amount of coherent current on the sample plane inside a FE-TEM. A methodological one, optimizing the optical conditions of the instrument in order to take maximum advantage of a specific source defined by a given brightness B. And an instrumental direction, optimizing directly the brightness B of the source by improving some key components. Indeed, following Eq. (1.37), improving B linearly increases the amount of coherent current. These two directions represent the skeletons of my research work and my future project.

In the following, I will first start with the description of my major methodological developments implemented using standard FE-TEM instruments such as the SACTEM-Toulouse, an FEI TECNAIF20 equipped with a C-COR system, and the In situ Interferometry TEM (I2TEM), an Hitachi HF3300 equipped with a B-COR system. I will then continue in Chapter 2 with the description of my instrumental works which concern the development of high brightness and ultrafast CFEG paving the way of coherent ultrafast TEM experiments.

1.2. Structural analysis using FETEM

During the first years of my research work, I was mainly interested in developments of coherent TEM methods allowing to determine quantitatively crystal strain state at the nanoscale. Target applications were semiconductor or optoelectronic devices, as an increasing number of these devices used stress engineering technologies requiring sensitive and local strain measurement methods.

I first developed diffraction and holography based methods on model epitaxial systems. Indeed, epitaxial thin layers undergo an elastic strain, related to the difference in lattice parameters, sometimes associated to chemical segregation. These various effects have a direct impact on the epitaxial thin layer properties (emission in quantum wells, transport and magnetic anisotropy...) (Hanbücken and Deville, 2001). They interest both fundamental research and applications aiming at adjusting the properties of epitaxial layers through the control of their strain state. Additionally, these techniques have been implemented on various kind of systems requiring local and sensitive strain informations (e.g., implanted crystals, real electronic devices, plastically strained crystal, ...). In the following I will systematically select the most relevant studies with regards to the contributions of coherent TEM and not from the material questions point of view.

Most of the experimental techniques used to determine the lattice parameters only provide informations averaged over the whole layer or its surface (De Wolf et al., 2003). On the contrary, TEM makes it possible to precisely select the studied zones, which is a considerable advantage in the presence of heterogeneities. In particular very accurate measurements can be obtained using highly localized CBED patterns obtained thanks to the use FE-TEM (Morniroli, 2002). The high sensitivity of this method is however weakened by the need of dynamical simulations of the experimental diffraction patterns to be able to properly retrieve the real crystal strain state (Houdellier et al., 2008a; Spence and Zuo, 2013).

Following the pioneering works of Hÿtch et al. in the use of geometrical phase analysis (GPA) to determine localized strain informations inside HREM micrographs of crystals or devices (Hÿtch et al., 1998), I have also worked on the development of dark-field electron holography (DFEH) allowing to extract strain state informations more easily than using CBED and without sacrificing accuracy (Hÿtch et al., 2008a).

All these approaches have benefited considerably from the design of the "I2TEM" performed in collaboration with the Hitachi company. Especially, I was interested in the implementation of split beam methods, either using CBED and holography configurations, to improve their potential for strain analysis. Split beam capabilities of I2TEM are indeed unique (Houdellier et al., 2015b; Röder et al., 2016).

1.2.1 Local strain measurement using CBED

1.2.1.1 Introduction to CBED geometry and diffraction theory

A schematic diagram of the CBED configuration is given in Fig. 1.9A. An incident convergent cone of electrons (in blue) illuminates a crystalline sample. Undiffracted beams (reported also in blue) continue in straight lines through the specimen, are then deflected by the objective lens to form a disk in its BFP and finally form the image of the spot in the image plane. Each direction of propagation inside the cone is considered independent of other

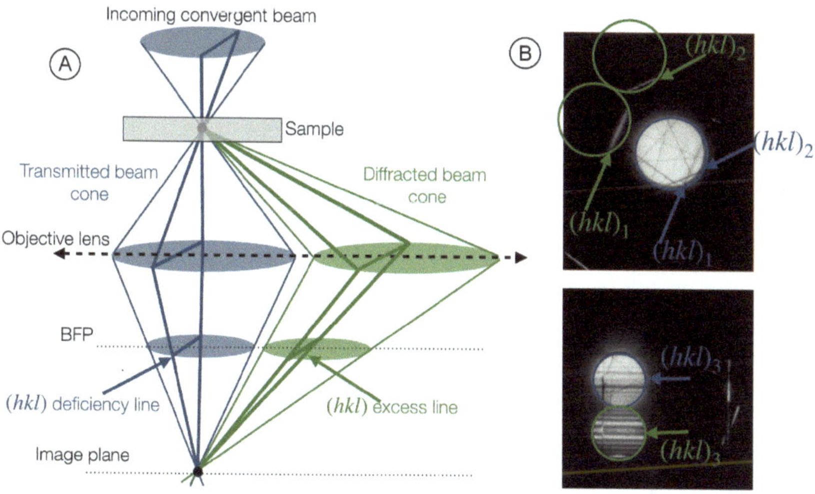

Fig. 1.9 (A) Optical outline describing the CBED configuration. The drawing is simplified compare to the real setup considering only the effect of the imaging objective lens. BFP stands for back focal plane. (B) Typical CBED patterns observed using a silicon sample. The *top* pattern has been recorded out of a main zone axis orientation showing mainly HOLZ kinematical lines. The *bottom* one is oriented in dynamical two-beam condition showing the presence of a dynamical ZOLZ line.

directions. The convergent spot formation mechanism in the object plane is the same as the one described for coherent STEM imaging in Section 1.1.3.1. Because the spot is located in the same optical plane as the sample, it should be noted that no image information will be observed in the convergent beam pattern. The CBED pattern is therefore composed of disks where each point within a disk corresponds to a specific direction of incidence but the same specimen thickness. The diameter of the individual disks is mainly controlled by the choice of the condenser aperture and may be adjusted (changing the convergence angle) by changing this aperture (e.g., to prevent overlap of diffraction disks) (Morniroli, 2002).

As reported in Fig. 1.9A, when a set of directions inside the incident cone are oriented in Bragg condition with a set of lattice plane (*hkl*) inside the crystal, these directions will be diffracted in the diffracted disk (in green). Going back to the objective lens BFP, a black deficiency line of electrons will be observed inside the transmitted disk and a white excess line of electrons in the diffracted disk of the CBED pattern. These two lines will be separated by an angle $2 \times \theta_B$ where θ_B corresponds to the Bragg angle calculated using the Bragg law and the lattice distance d_{hkl}: $2d_{hkl}\sin(\theta_B) = \lambda_0$.

Perpendicular to the diffracted lines in the incident cone, we find incident electron directions, defined by an excitation error $s_{\vec{g}}$ relative to the directions in exact Bragg condition with (*hkl*) lattice planes (Morniroli, 2002). Naturally any set of lattice planes (*hkl*) could be described in the reciprocal space in terms of vector $\vec{g}$ where $|\vec{g}| = 1/d_{hkl}$. The change in the excitation error, as well as the direction of change, across the CBED disk is thus important for understanding the rich diffraction intensity patterns often observed in CBED. Indeed, the intensity profile taken perpendicular to all diffracted lines associated to a lattice plane (*hkl*) will correspond to the evolution of diffracted intensities as a function of the excitation error $I_{\vec{g}}(s_{\vec{g}})$ and is known as a rocking curve (Spence and Zuo, 2013).

Fig. 1.9B reports the two kinds of rocking curves observed in standard CBED patterns. Some diffracted lines will exhibit very simple profiles even close to a perfect line, with a clear maximum for the excess line (or a minimum for the deficiency line). This maximum is observed for directions inside the incident cone satisfying the excitation error $s_{\vec{g}} = 0$, which corresponds to the exact Bragg position (e.g., $(hkl)_2$ in Fig. 1.9B). More complex lines could also be observed displaying many fringes in the rocking curve. The first type of rocking curves can be simulated using the kinematical approach of the electron diffraction theory. Such as for crystal X-ray diffraction, kinematical theory calculates diffracted intensities using crystal structure factors $F_{\vec{g}} = A\sum_i f_a(s)\exp(-B_i s^2)\exp(-2\pi i\vec{g}.\vec{r}_i)$ where A is a constant linked to Ω the unit cell volume, $s = |\vec{g}|/2$ and B_i corresponds to the Debye–Waller factors (Spence and Zuo, 2013). They can be determined using electron atomic scattering factors $f_a(s)$ and knowing the atom positions $\vec{r}_i$ thanks to the crystal symmetry. Using the kinematical approximation, the diffracted intensities $I_{\vec{g}}(s_{\vec{g}})$ can be written (Hirsch et al., 1977; Spence and Zuo, 2013):

$$I_{\vec{g}}(s_{\vec{g}}) = \frac{\pi^2}{\xi_g^2}\frac{\sin^2(\pi t s_{\vec{g}})}{(\pi s_{\vec{g}})^2} \quad (1.44)$$

where t is the crystal thickness, and $\xi_{\vec{g}}$ the extinction distance connected to the structure factor for a reciprocal lattice point $\vec{g}$. The extinction distance is then a fundamental property of the diffracted intensity $\vec{g}$ and can be interpreted as the minimum sample thickness from which the diffracted beam will be diffracted again by the set of lattice planes $\vec{g}$. The kinematical theory is valid only for very thin objects for which the reflection intensity $I_{\vec{g}}(s_{\vec{g}})$

is small and the decrease of the primary-beam intensity I_0 can be neglected. It is commonly admitted that kinematical theory is valid when $I_{\vec{g}}(s_{\vec{g}}) \ll 1$ given by $t < \xi_{\vec{g}}/10$ or $s_{\vec{g}} \gg 1/\xi_{\vec{g}}$. Using Eq. (1.44) we find a maximum of the kinematical diffracted intensity at $s_{\vec{g}} = 0$ corresponding the line position in Fig. 1.9B (Morniroli, 2002). Furthermore, kinematical theory considered the excitation of only one Bragg reflection at a time. This condition is known as the two-beam case and is unusual; normally, a larger number $N > 2$ of reflections must be considered. We are then in the N-beams case which cannot be described properly by the kinematical theory. In fact, the N-beam condition restricts the validity of the kinematical theory to even smaller thicknesses.

A more rigorous description of electron diffraction by crystalline materials has to take into account multiple scattering, known as dynamical diffraction effects (Zuo, 2017). This description is called dynamical theory of the electron diffraction, and can be used to characterize quantitatively, with a surprising accuracy, the diffracted rocking curve profiles obtained either in two-beam or N-beam conditions (Spence and Zuo, 2013). We will now quickly introduce the main ideas behind dynamical theory.

Outside the crystal, it is natural to think of the transmitted and diffracted waves emerging from a specimen as plane waves. The exit wave is then described by the sum of all these emerging waves with appropriate phase factors $e^{i\left(\vec{k}+\vec{g}\right).\vec{r}}$, where $\vec{k}$ is the incident wave vector and $\vec{g}$ the associated reciprocal lattice vector. However inside the crystal it is often more convenient and useful to consider the excitation of Bloch waves which correspond to waves that have the periodicity of the lattice (Zuo, 2017). Each Bloch wave is defined by an excitation amplitude written C^j for the jth Bloch wave and a wave vector $\vec{k^j}$. When $\vec{k^j} = \vec{k}$, there will be virtually no diffraction. In general, each $\vec{k^j}$ maps out a surface in the reciprocal space known as the dispersion surface. The dispersion surface of accelerated electrons crossing a crystal is equivalent of the Fermi surface for conduction electrons. We associate a separate branch of the dispersion surface for each Bloch wave j. In the kinematical approximation this dispersion surface would consist a series of spheres of radius $\vec{k}$ centered on each reciprocal lattice point. But due to dynamical effects, each branch of the dispersion surface will move away from the so-called kinematical spheres. Each Bloch wave has a different spatial distribution in a crystal lattice and therefore a different potential energy, but because the main concern is elastic scattering when the potential energy decreases, the kinetic energy, associated with $\vec{k}^j$ increases and vice versa, so that energy is conserved. The total energy of one bloch wave is noted γ_j and is often called the anpassung in the literature (Spence and Zuo, 2013; Zuo, 2017).

Let's briefly described the calculation of diffracted intensities using dynamical theory. Starting in two-beam conditions, the electron propagation inside the crystal will be described using two Bloch states. These two Bloch states will be located in the reciprocal space along two separate branches of the dispersion surface defined by their own energy γ_1 and γ_2 as drawn in Fig. 1.10.

Mathematically, in two-beam condition, each Bloch wave is described by the sum of two plane wavelets (Zuo, 2017):

$$\begin{aligned}\Phi_1(\vec{r}) &= C_0^1 e^{2\pi i \vec{k}_0^1 . \vec{r}} + C_{\vec{g}}^1 e^{2\pi i \vec{k}_{\vec{g}}^1 . \vec{r}} \\ \Phi_2(\vec{r}) &= C_0^2 e^{2\pi i \vec{k}_0^2 . \vec{r}} + C_{\vec{g}}^2 e^{2\pi i \vec{k}_{\vec{g}}^2 . \vec{r}}\end{aligned} \tag{1.45}$$

where $\vec{k}_0^{1,2}$ and $\vec{k}_{\vec{g}}^{1,2} = \vec{k}_0^{1,2} + \vec{g}$ are determined by the intersection between $\vec{n}$, the vector normal to the sample surface, and the two branches of the

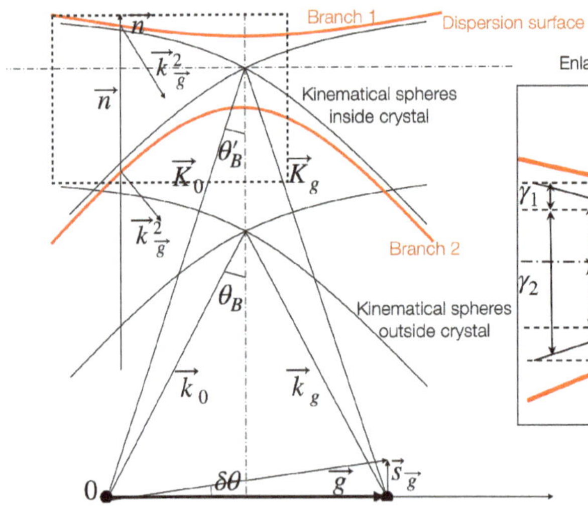

Fig. 1.10 Description of the formation of an electron diffraction pattern using Bloch wave dispersion surfaces drawn in the crystal reciprocal space. Kinematical spheres inside and outside the material, showing the refraction effect due to the mean inner potential, are also reported.

dispersion surface (see Fig. 1.10). These wave-vectors implicitly contain information about the Bloch state energy γ_1 and γ_2 as $\vec{k^j} = \vec{K} + \gamma_j \vec{n}$ where $\vec{K}$ is the wave-vector inside the crystal described by the kinematical sphere (i.e., considering plane waves). The transmitted exit wave can be easily reconstructed combining the two transmitted wavelets $\vec{k_0^1}$ and $\vec{k_0^2}$ coming from the two Bloch states following the expression: $\Psi_0 = c_1 C_0^1 e^{2\pi i \vec{k_0^1}.\vec{r}} + c_2 C_0^2 e^{2\pi i \vec{k_0^2}.\vec{r}}$. The excitation coefficients c_j are to be determined by matching the incident waves with waves inside the crystal at the entrance surface. And the same operation is performed for the diffracted beam using the two wavelets $\vec{k_{\vec{g}}^1}$ and $\vec{k_{\vec{g}}^2}$ defined in the two Bloch states. Knowing the two transmitted and diffracted wave-functions, the associated intensities could then be calculated very easily using $I_0 = |\Psi_0|^2$ and $I_{\vec{g}} = \left|\Psi_{\vec{g}}\right|^2$.

It is now possible to generalize the previous description in N-beam (Spence and Zuo, 2013). Indeed, in N-beam configuration the electron propagation inside the crystal is defined by N Bloch states. The dispersion surface will then exhibit N branches. As a consequence, each Bloch wave j will be composed by the sum of N wavelets $C_0^j, C_{\vec{g}_1}^j, C_{\vec{g}_2}^j, \ldots C_{\vec{g}_N}^j$. As for the two beams condition, the transmitted/diffracted wave function in N beams configuration will be calculated by combining the N transmitted/diffracted wavelets. The total wave function is then obtained by summing the transmitted and diffracted wave-functions (Spence and Zuo, 2013):

$$\Psi\left(\vec{k}, \vec{r}\right) = \sum_j c_j \sum_{\vec{g}} C_{\vec{g}}^j e^{2\pi i \gamma_j t} e^{2\pi i \left(\vec{k}_0 + \vec{g}\right).\vec{r}} = \sum_{\vec{g}} \psi_{\vec{g}} e^{2\pi i \left(\vec{k}_0 + \vec{g}\right).\vec{r}} \tag{1.46}$$

where $C_{\vec{g}}^j$ are the Bloch wave coefficients and t is the specimen thickness. If we consider the surface normal $\vec{n}$ approximatively antiparallel to the beam, we can show that the excitations coefficients c_j are equal to $(C_0^j)^{-1}$. The complex amplitude of the wave function of the diffracted beam $\vec{g}$ can be written as:

$$\psi_{\vec{g}} = \sum_{j=1}^{n} \left(C_0^j\right)^{-1} C_{\vec{g}}^j e^{2\pi i \gamma_j t} \tag{1.47}$$

and the transmitted beam 0 can be written as follows:

$$\psi_0 = \sum_{j=1}^{n} \left(C_0^j\right)^{-1} C_0^j e^{2\pi i \gamma_j t} \tag{1.48}$$

In order to establish how these quantities depend on the crystal nature, we have to determine the Bloch wave coefficients $C_{\vec{g}}^j$ and energy γ_j as a function of the structure factor $F_{\vec{g}}$. The inner potential of a given crystal can be expanded as Fourier series $U(\vec{r}) = \sum_{\vec{g}} U_{\vec{g}} e^{2\pi i \vec{g}.\vec{r}}$, where $U_{\vec{g}}$ are the complex Fourier components. These components are linked to the structure factor $F_{\vec{g}}$ or the extinction distance $\xi_{\vec{g}}$. Inserting the wave function expression $\Psi(\vec{k}, \vec{r})$, given by Eq. (1.46), into the Schrödinger equation describing the propagation of the electron wave into the crystal periodic potential $U(\vec{r})$, we obtain the so-called secular equation (Spence and Zuo, 2013). This equation can be solved numerically to obtain Bloch wave coefficients $C_{\vec{g}}^j$ and the anpassungen γ_j for any kind of crystal defined by their sets of structure factor. It can be written in the N-beam condition as follows:

$$\frac{B_{\vec{g}}^j\left(2\vec{K}.\vec{g} + g^2\right)}{1 + g_n/K_n} - \sum_{\vec{h}} \frac{B_{\vec{h}}^j U_{\vec{g}-\vec{h}}}{(1 + g_n/K_n)^{1/2}(1 + h_n/K_n)^{1/2}} = -2K_n\gamma_j B_{\vec{g}}^j \tag{1.49}$$

where $\vec{K}$ is the wave vector inside the crystal, $g_n = \vec{g}.\vec{n}$, $K_n = \vec{K}.\vec{n}$ and

$$B_{\vec{g}}^j = \left(1 + \frac{g_n}{K_n}\right)^{1/2} C_{\vec{g}}^j \tag{1.50}$$

The effect of absorption is included through the addition of a small imaginary component for each Fourier component of the potential $U_{\vec{g}}$ (Zuo, 2017).

Fig. 1.11 shows a comparison between an experimental CBED pattern and two simulations performed using either the kinematical approximation or N-beam dynamical simulation (Houdellier, 2006). This pattern is the central transmitted disk of a CBED pattern observed at $V_0 = 200\,\text{kV}$ in a silicon crystal oriented in [310] zone axis orientation. In the case of kinematical simulation, only the maximum of the rocking curve is displayed showing sharp fine lines whatever the reflection $\vec{g}$. This comparison clearly demonstrates that dynamical theory is the only approach which could reproduce all the intensity variations across the CBED disk.

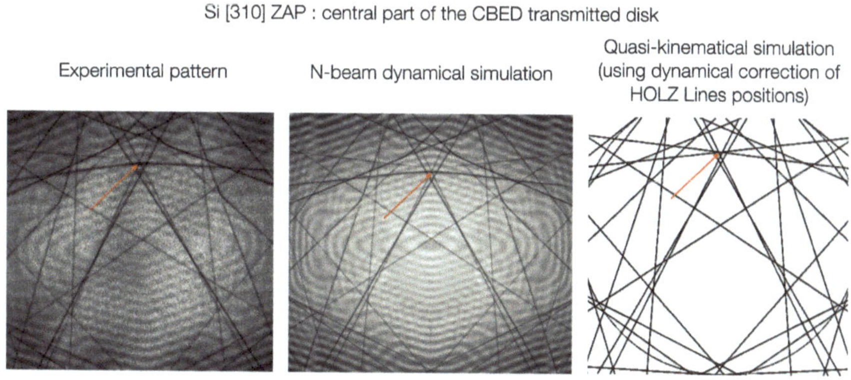

Fig. 1.11 Central disk of a [310] oriented CBED pattern recorded using a silicon sample. The acquisition was performed at 200 kV. The HOLZ lines positions can be compared with those obtained using N-beam dynamical simulation and a quasi-kinematical simulation. *Extracted from Houdellier, F., 2006. Contribution au développement du CBED et de l'holographie HREM pour l'analyse des déformations de couches épitaxiées. PhD thesis INSA Toulouse and CNRS-CEMES (2006). INSA, Toulouse, PhD thesis.*

If the optic axis is lined up with a crystal zone axis, then a particularly convenient CBED pattern is obtained, often called a zone axis pattern or ZAP (Morniroli, 2002). A zone axis is the normal to a plane in the reciprocal lattice. The $<uvw>$ zone axis is a line common to all crystal planes (hkl) that obey the equation $uh + vk + wl = 0$. It is particularly helpful to describe the geometry of a diffraction pattern with the Ewald sphere construction within the reciprocal lattice (Morniroli, 2002). When the electron beam is incident along a zone axis direction, the Ewald sphere makes a near planar coincidence with a plane called the zero-order Laue zone (ZOLZ) of the reciprocal lattice. With increasing distance from the zone axis, the Ewald sphere moves away from near coincidence with the ZOLZ, and the reflections are no longer excited until the Ewald sphere intersects the next layer of the reciprocal lattice parallel to the ZOLZ. A large-diameter circle is obtained with the intersection of the Ewald sphere and the next-layer plane called first-order Laue zone (FOLZ). Usually, all the layers above the ZOLZ are called high-order Laue zone (HOLZ).

The simple geometrical construction reported in Fig. 1.9A finally corresponds to a two-beam kinematical description of the CBED pattern formation. We can easily generalize this simple geometrical approach to the case of many beams. In this case the transmitted disk of the CBED pattern will exhibit the superposition of all deficiency lines associated with the N-1 diffracted excess lines distributed around it. This is basically what we can observe in the first pattern of Fig. 1.9B and the simulations of Fig. 1.11. Of course this approach doesn't take into account the dynamical effects, but is sufficient to describe geometrically the distribution of lines in the CBED transmitted disk. Some lines in the CBED pattern will originate from HOLZ diffraction. They are called HOLZ lines. On the contrary, ZOLZ lines will come from the ZOLZ diffractions (Morniroli, 2002). In two-beam condition, HOLZ lines usually exhibit fine profiles, which can be simulated using kinematical approach as their associated extinction distances $\xi_{\vec{g}}$ are larges. On the other hand, ZOLZ lines will require a dynamical approach to reproduce their rocking curves even in two-beam condition. To reproduce all the intensity variation observed in N-beam condition, even considering a simple superposition of HOLZ lines, we will need dynamical simulations. See for instance the dynamical effect observed in the HOLZ lines intersection, highlighted using a red arrow in Fig. 1.11, reproduced by the dynamical simulation but not by the kinematical one (Spence and Zuo, 2013).

1.2.1.2 Strain state determination in uniform crystal

Crystal lattice parameters can be measured using the Bragg law in diffraction experiments. They can be measured very precisely from HOLZ line positions in CBED as these lines are sharp and then provide very good signal over noise ratio for the determination of small lattice parameters variation (Zuo, 1992). Furthermore, HOLZ lines with higher indices (*hkl*) can be observed in the transmitted disk. Indeed, they have also higher sensitivity to small variations of lattice parameters compare to ZOLZ reflections. In practice, the experimental HOLZ line positions have to be compared to simulations through the Hough transformation procedure (Leavers, 1992). As already described previously, the ideal situation should be to compare the HOLZ lines positions with dynamical simulations. Fig. 1.12 reports an extreme application of this method selected for it's high sensitivity (Boninelli et al., 2018). Indeed, in this study a structural investigation of P doped germanium sample has been carried out. Doping was performed by ion implantation and laser thermal annealed was then realized to recrystallize and activate amorphized germanium doped areas with phosphorous. Laser thermal annealing (LTA) was carried out at various energy densities. The purpose of this study was to investigate how the damage recovery occurs as a function of the LTA conditions. Indeed, despite phosphorus exhibiting a significantly smaller atomic radius compared to Ge, the theory predicts that its inclusion in Ge should generate an expansion of the lattice due to

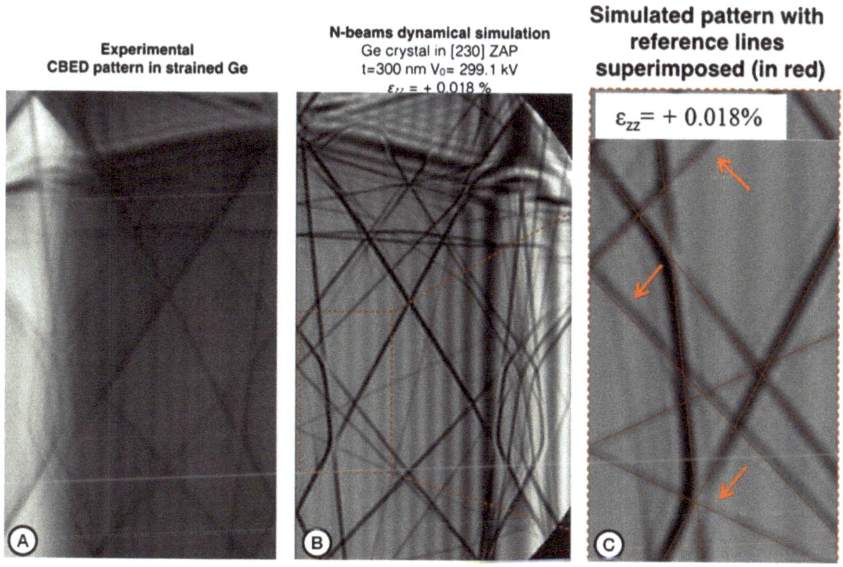

Fig. 1.12 Experimental (A) and simulated (B) CBED pattern taken at ~50 nm from the top sample surface in the doped region. The comparison between the N-beams dynamical simulated CBED patterns of the strained area (in *gray*) with the unstrained reference (in *red*) are reported in (C). The shift of HOLZ line positions, marked by *arrows*, is ascribed to a variation of lattice parameter. *Extracted from Boninelli, S., Milazzo, R., Carles, R., Houdellier, F., Duffy, R., Huet, K., La Magna, A., Napolitani, E., Cristiano, F., 2018. Nanoscale measurements of phosphorous-induced lattice expansion in nanosecond laser annealed germanium. APL Mater. 6 (5), 058504.*

complex mix between the electronic and the size contribution (Boninelli et al., 2018). This effect is theoretically very small (below 10^{-4} of relative displacement) and was never clearly demonstrated. I choose this study among others as it clearly demonstrates CBED capability to resolve structural questions requiring over-sensitivity.

For this experiment, CBED measurements were conducted in STEM mode allowing to scan the beam and acquire CBED patterns for each spot position taken in the doped region located beneath the top sample surface and running over a depth of approximatively 100 nm. Localized strain measurements could then be achieved with nanometer spatial resolution. N-beams dynamical simulations were performed using Bloch wave approach described previously and computed using JEMS software (Stadelmann, 1987). A homemade Hough transform code was implemented in Gatan Digital Micrograph software to extract the HOLZ lines positions (Leavers, 1992). All the experimental CBED patterns were acquired in the [230] zone axis using the Hitachi I2TEM microscope and were zero loss filtered with a 10 eV energy window to remove inelastic electrons bringing unnecessary noise to the measurement. In addition to the lattice parameters, HOLZ line positions are also very sensitive to the TEM accelerating voltage (Morniroli, 2002). Therefore, a CBED pattern obtained in the undoped (and undistorted) substrate (located further away than 100 nm below the surface) was initially simulated to determine the precise acceleration voltage of the electron beam. It was found equal to $V_0 = 299.1\,\text{kV}$ and then used as a reference for the strain measurements. In CBED, the perpendicular strain in the doped region $\varepsilon = (a_{DGe} - a_{Ge})/(a_{Ge})$ is determined by measuring the crystal lattice parameter in the doped region a_{DGe} with respect to the undoped Ge substrate lattice parameter a_{Ge}. Measurements of lattice parameters performed on the fully activated structures show that P doping results indeed in a lattice expansion, with a perpendicular lattice strain per atom $+0.7 \pm 0.1\,\text{Å}^3$ in a good agreement with previous theoretical predictions (Boninelli et al., 2018).

Dynamical simulation is ideal to implement this kind of comparison with experimental CBED patterns. But it requires a full calculation of the ZOLZ dispersion surface and, depending on the zone axis orientation, the number of reflections which have to be considered could lead to highly time consuming calculations. Rather, kinematical simulations can be obtained in less than *ms* with any computer, but the major factor limiting factor comes from the strong influence of dynamical effects (Stadelmann, 1987).

In any crystal it is however possible to find specific zone axis orientation where only one branch of the ZOLZ dispersion surface is excited. These orientation are called quasi-kinematical zone axis. In that case, dynamical effects which cause the dispersion surface to deviate from its kinematical location leading to wrong HOLZ line positions in kinematical simulations, can be corrected. Indeed, considering a single HOLZ reflection, the ZOLZ dispersion surface effect on this HOLZ line position can be simulated by adjusting the high voltage in the kinematical simulation. This correction called HOLZ shift, enables to simulate correct HOLZ lines positions using kinematical simulations preventing the use of complex dynamical simulations. Kinematical simulations, with HOLZ shift correction which includes dynamical effects on the HOLZ lines positions, are called quasi-kinematical simulations and are only valid along quasi-kinematical zone axis (Bird, 1989). We used this method to determine the evolution of strain over large field inside Si crystal after plastic deformation (Legros et al., 2008). This kind of study requires many CBED patterns to be compared, and the use of quasi-kinematical simulations was very helpful.

1.2.1.3 Measurements in the presence of inhomogeneous crystal strain

The HOLZ line profile depends also sensitively on strain variations and crystal rotations inside the crystal volume probed by the electron beam. The previous method used to determine lattice parameters thanks to HOLZ lines position can only be applied if fine HOLZ lines are observed. This is possible only if the lattice parameter is uniform along the electron beam path such as for perfect crystal or uniformly strained crystal. The last case can be encountered in weakly strained samples.

A highly strained sample will exhibits surface strain relaxation during the milling operation needed to prepare the TEM sample (Clément et al., 2004). The relaxation leads to strain variations across the sample thickness, and HOLZ line rocking curve broadening. This effect was observed very quickly during the analysis of *SiGe* epitaxial layer deposited on *Si* substrate (Houdellier, 2006). Due to surface strain relaxation, CBED patterns recorded in the substrate, even far away from the epitaxial interface, exhibit impressive HOLZ lines broadening. This measurement is reported in Fig. 1.13, where we can notice that HOLZ broadening increase the closer we are from the epitaxial interface. The sample was prepared by FIB using a specific geometry reported in Fig. 1.13 and the CBED patterns were acquired using the SACTEM-Toulouse and were zero loss filtered with a 10 eV energy window. Specimen was oriented along a [230] zone axis.

The effect of a deformed lattice on the HOLZ line rocking curve can be qualitatively understood using the kinematical diffraction theory of deformed crystals where a z dependent displacement field $\vec{u}(z)$ can be inserted in the z dependent diffracted plane wave $\Psi_{\vec{g}}(z)$ by adding a phase factor $\exp(-2\pi i\vec{g}.\vec{u}(z))$ (Clément et al., 2004). The final diffracted intensity can be calculated in the form of an integration over the crystal thickness t along the incident beam direction z : $\left|\int_0^t \Psi_{\vec{g}}(z)dz\right|^2$. It shows that the recorded broadening is far greater for a HOLZ reflection than for a low-order reflection. Indeed, the phase factor will increase with $|\vec{g}|$ which makes the HOLZ line profile very sensitive to $\vec{u}(z)$. On the contrary, we can see that ZOLZ reflections rocking curve will not be sensitive to $\vec{u}(z)$.

Carrying out quantitative measurement is also possible by comparing experimental patterns with dynamical simulations taking into account the displacement field $\vec{u}(z)$. These simulations have to be associated with finite element modeling used to calculate $\vec{u}(z)$ (Houdellier et al., 2006). In order to introduce the effect of z dependent displacement field in the formalism of dynamical theory, we have implemented a perturbation approach, which is

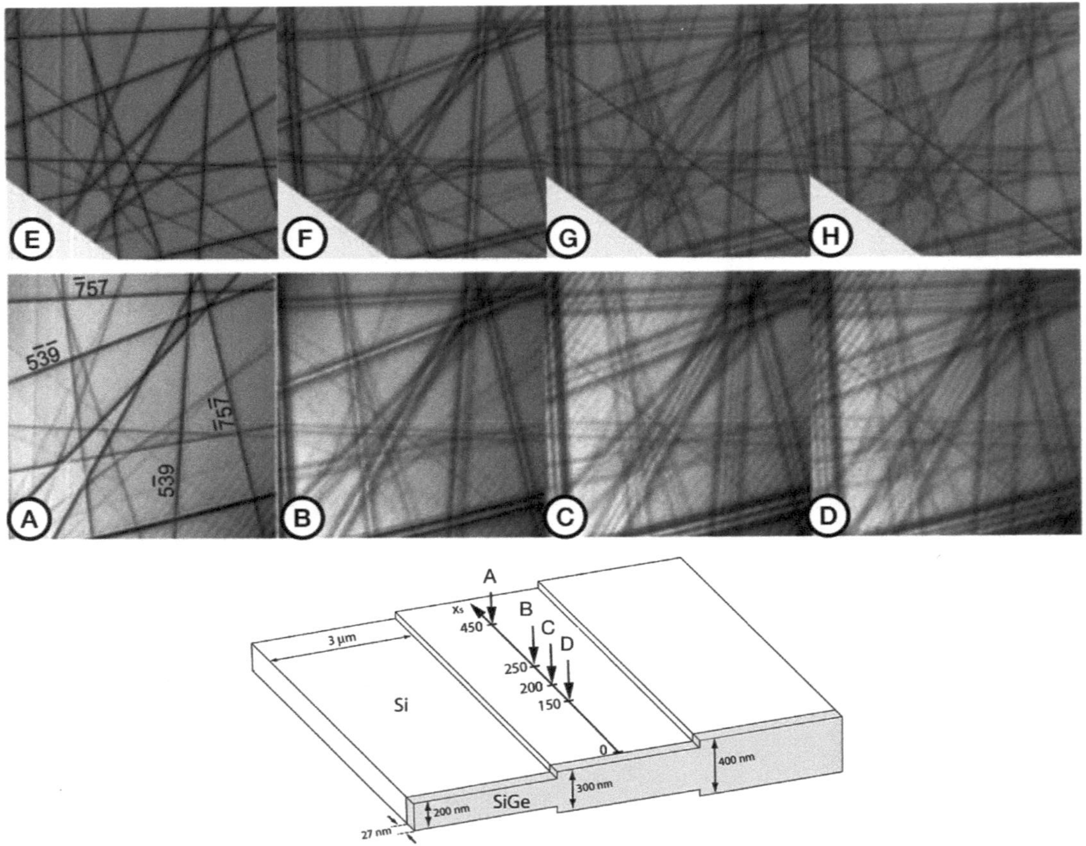

Fig. 1.13 [230] CBED central disks of a $t = 300\,nm$ thick cross-sectional $Si_{0.8}Ge_{0.2}/Si$ specimen taken in the silicon region at respective distances (from *left* to *right*) of $x_s = 450$, 250, 200 and 150 nm from the interface with the epitaxial layer. *Bottom*: experimental patterns; *top*: dynamical simulated patterns using perturbation approach. The scheme below shows the geometry of H-bar FIB prepared sample. Measurements were performed on the central thickness at 200 kV. *Extracted from Houdellier, F., 2006. Contribution au développement du CBED et de l'holographie HREM pour l'analyse des déformations de couches épitaxiées. PhD thesis INSA Toulouse and CNRS-CEMES (2006). INSA, Toulouse, PhD thesis.*

finally very close to the eikonal method used in charged particles optics to determine the aberrations coefficients (Houdellier, 2006; Houdellier et al., 2008a). Indeed in CPO, as introduced in the beginning of this chapter, second-order characteristic function will allowed to determine paraxial properties such as all the cardinal elements. Aberrations, which can be seen as a z dependent perturbations of the paraxial trajectories, can be simulated by considering the fourth-order contribution of the characteristic function as a perturbation of the paraxial one (Rose, 2009).

To apply this method, we will first determine the Hamiltonian function of the electron inside the crystal $H(\vec{r})$. We could then define the associated evolution operator $\mathcal{U}(0,\ z) = \exp\left(-i\int_0^z H(\tau)d\tau\right)$ (Gratias and Portier, 1983), and considered it as a characteristic function of the system in the sense of Hamilton. The idea will be to express the evolution operator $\mathcal{U}(0,\ z)$ into a perturbation series. Low-order term of this expansion will be defined by the ZOLZ potentials, and higher one by the HOLZ and inhomogeneous displacement field contributions. Just as in the case of the eikonal method, the contribution of the higher order will be treated as a perturbation of the low-order characteristic function.

Performing this separation requires the reciprocal lattice vector to be written as $\vec{g} = \left(\vec{G}, -ng_z\right)$, where $\vec{G}$belongs to the ZOLZ, g_z is the distance between two neighboring Laue zones and n the order of the zone. Using these notations, the crystal potential $U(\vec{r})$ can then be developed as (Houdellier et al., 2008a):

$$U(\vec{r}) = \sum_g U_{\vec{g}} e^{2\pi i\vec{g}.\vec{r}} = U^{(0)}(\vec{\rho}) + \sum_{n\neq 0} \exp(-2\pi i n g_z z)U^{(n)}(\vec{\rho}) \quad (1.51)$$

in which $\vec{\rho}$is the two-dimensional vector corresponding to the projection of $\vec{r}$ in the $(x,\ y)$ plane and $U^{(n)}(\vec{\rho}) = \sum_{\vec{G}} U^{(n)}_{\vec{G},-ng_z} \exp\left(2\pi i\vec{G}.\vec{\rho}\right)$. In this expression, the ZOLZ potential $U^{(0)}(\vec{\rho})$ corresponds to the z independent ZOLZ projected potential while the other components $U^{(n)}(\vec{\rho})$ are known as the higher-order conditional projected potentials and correspond to the HOLZ contribution (Houdellier et al., 2008a).

We can do the same operation introducing an inhomogeneous displacement field $\vec{u}(z)$ modifying the original atom position $\vec{r}$ to it's new z-dependent position $\vec{r}' = \vec{r} - \vec{u}(z)$. The resulting strained crystal potential $U(\vec{r})$ can be developed as:

$$U(\vec{r}) = \sum_{\vec{g}} U_{\vec{g}} e^{2\pi i \vec{g}.(\vec{r}-\vec{u}(z))} = \sum_{\vec{g}} U_{\vec{g}} e^{2\pi i \vec{g}.\vec{r}} + \sum_{\vec{g}} U_{\vec{g}} e^{2\pi i \vec{g}.\vec{r}} \left[\exp(-2\pi i \vec{g}.\vec{u}(z) - 1\right]$$
$$= U^{(0)}(\vec{\rho}) + U^{\mathrm{HOLZ}}(\vec{\rho}, z) + U^{\mathrm{strain}}(\vec{\rho}, z) = U^{(0)}(\vec{\rho}) + U^{\mathrm{pert}}(\vec{\rho}, z) \tag{1.52}$$

Once again, the strained crystal potential can be separated into a z-independent term $U^{(0)}(\vec{\rho})$ and a z-dependent one, as performed in Eq. (1.51) for the perfect crystal. Note that the z-dependent term $U^{\mathrm{pert}}(\vec{\rho}, z)$ now includes both the HOLZ potential $U^{\mathrm{HOLZ}}(\vec{\rho}, z)$ and the inhomogeneous displacement field $U^{\mathrm{strain}}(\vec{\rho}, z)$ contributions.

Introducing Eq. (1.52) into the Hamiltonian $H(\vec{r})$ and taking care of the small angle approximation, allowed us to separate the Hamiltonian function in two parts, one z-independent and one z-dependent. Finally $H(\vec{r})$ can be written as:

$$2k_z H(\vec{r}) = 2k_z H^{0=\mathrm{ZOLZ}}(\vec{\rho}) + U^{\mathrm{pert}}(\vec{\rho}, z) \tag{1.53}$$

Standard dynamical diffraction theory, outlined in Section 1.2.1.1, can now be applied by simply considering the Hamiltonian $H^{0=\mathrm{ZOLZ}}(\vec{\rho})$ instead of $H(\vec{r})$. Indeed, this theory was described for a perfect three-dimensional crystal. In that case three-dimensional Bloch waves are the eigenstates of $H(\vec{r})$ because the potential $U(\vec{r})$ is perfectly periodic over the three spatial dimensions (Spence and Zuo, 2013). They are sufficient to fully calculate all the diffracted intensities. In that case implementing the perturbation theory considering the HOLZ potential as a perturbation of the ZOLZ diffraction will give the same results but is not of particular interest (Gratias and Portier, 1983). However the perturbation method becomes attractive considering the inhomogeneous displacement field.

Considering $H^{0=\mathrm{ZOLZ}}(\vec{\rho})$ two-dimensional Bloch waves $|j>$ are now the new eigenstates. Dirac notations $|>$ are used here to simplify the following expressions. However, classical wave-functions can be retrieved very simply by projecting the state on the correct basis; e.g., let's note the first three-dimensional Bloch eigenstate $|1>$, solution of the dynamical two beams diffraction; the associated wavefunction, defined in Eq. (1.45), is then calculated by its projection on the real space basis $|\vec{r}>$ using $<\vec{r}|1>= \Phi_1(\vec{r})$. Going back to the ZOLZ potential, we can now write:

$$H^{0=\mathrm{ZOLZ}}(\vec{\rho})\,|j>= \gamma_j|j> \tag{1.54}$$

where γ_j corresponds to the ampassungen of the two-dimensional Bloch waves as defined in Section 1.2.1.1. The associated evolution operator $\mathcal{U}^{(0}(0,z)$ can be written (Gratias and Portier, 1983):

$$\mathcal{U}^{(0}(0,z) = \sum_j e^{-i\gamma_j z}|j><j| \tag{1.55}$$

Knowing the incident electron beam plane wave state $|0>$, all the ZOLZ diffracted intensities, defined by their diffraction states $|\vec{q}>$, can be calculated introducing Eq. (1.55) into:

$$I_{\vec{q}} = |<0|\mathcal{U}^{(0}(0,z)|\vec{q}>|^2 \tag{1.56}$$

The description presented until now is strictly identical to the one presented in Section 1.2.1.1, restricting all the calculated intensities to the ZOLZ reflections.

Considering the potential $U^{\text{pert}}(\vec{\rho}, z)$ as a perturbation of the ZOLZ diffraction, the evolution operator can be expanded in a perturbation series (Houdellier, 2006; Houdellier et al., 2008a):

$$\mathcal{U}(0,z) = \mathcal{U}^{(0)}(0,z) + \sum_{m=1}^{0} \mathcal{U}^{(m)}(0,z) \tag{1.57}$$

with

$$\mathcal{U}^{(m)}(0,z) = -i\int_0^z d\tau \mathcal{U}^{(0)}(z,\tau)U^{\text{pert}}(\tau)\mathcal{U}^{(m-1)}(\tau,0) \tag{1.58}$$

Restricting the perturbation expansion to the first order is sufficient to reproduce all the fine diffracted contrasts observed in the transmitted beam of the CBED pattern, which can be calculated thanks to the following expression:

$$I_0 = |<0|\mathcal{U}^{(0}(0,z)|0> + <0|\mathcal{U}^{(1}(0,z)|0>|^2 \tag{1.59}$$

Results obtained with the perturbation approach were implemented using an homemade software coded in Fortran. They were compared with simulations given by standard dynamical theory described in Section 1.2.1.1 and computed using JEMS software, considering a perfect silicon crystal (Houdellier, 2006; Houdellier et al., 2008a). It shows the validity of the perturbation approach. Moreover, this method was also tested by simulating the HOLZ line splitting observed in LACBED around a dislocation due to it's

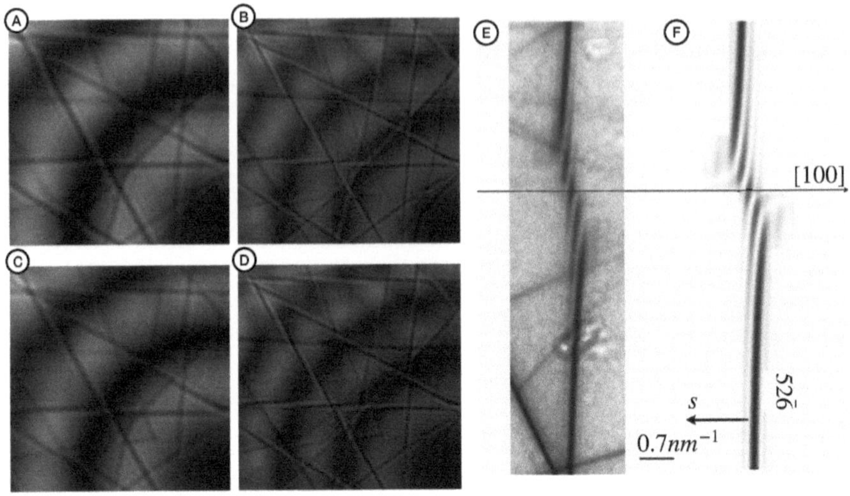

Fig. 1.14 Enlargements of the central region of [111] zone axis CBED patterns calculated for a nominal voltage of $V_0 = 200\,\text{kV}$ and a specimen thickness of $t = 350\,\text{nm}$ (A–C) and $t = 450\,\text{nm}$ (B and D). Comparison of HOLZ intensities calculated using standard dynamical theory (A and B) and perturbation theory (C and D). (E) Experimental and (F) simulated LACBED patterns showing the intersection of ($52\bar{6}$) HOLZ line by a dislocation line parallel to [100] direction and defined by a Burgers vector $\vec{b} = 1/2(01-1)$ in a $t = 400\,\text{nm}$ thick silicon crystal. *Extracted from Houdellier, F., Altibelli, A., Roucau, C., Casanove, M.J., 2008. New approach for the dynamical simulation of CBED patterns in heavily strained specimens. Ultramicroscopy 108 (5), 426–432.*

inhomogeneous strain field. Here too, these comparisons show the validity of the perturbation approach to deal with inhomogeneous strain field. These comparisons are reported in Fig. 1.14.

Finally, by calculating the field $\vec{u}(z)$ using the FEM software COMSOL multiphysics for the SiGe/si epitaxial system described in Fig. 1.13, and introducing this field in the simulation performed using the perturbation theory, we were able to compute all the experimental patterns reported in Fig. 1.13 (see in particular the simulations showed in Fig. 1.13E–H). The good agreement allowed us to determine the original strain state of the SiGe epitaxial layer as the initial condition introduced in the FEM simulations (Houdellier, 2006).

1.2.1.4 Summary on strain mapping using CBED

Thanks to FE-TEM it is possible to generate highly localized electrons spot focused on a sample. This spot can be scanned and CBED patterns can be acquired for each spot positions. Due to the high sensitivity of the HOLZ lines position to the lattice parameters, it is possible to determine very precisely a variation of these parameters (relative variation down to 10^{-4}) as long as these variations remain uniform over the electron propagation direction z.

However the major limitation of this HOLZ lines position fitting procedure for strain measurement came from the destructive process involved in the specimen preparation. First, and we didn't mention it before, stress and strain can be introduced by the new surfaces created during the sample preparation like for instance amorphous layer introduced during the polishing. But the main issue finally comes from strain relaxation when a 3D strained structure is prepared as a thin TEM lamella. This phenomena introduces an inhomogeneous displacement field along the electron propagation $\vec{u}(z)$ generating splitting of the HOLZ fine rocking curves. We showed that a possible solution to measure the strain state of a device should be to combine the experimental measurements with FEM simulations and use dynamical simulations to retrieve the original strain/stress. This method is schematically reported in Fig. 1.15. The 3D strain in the prepared cross-sectional sample could also be measured by comparing many HOLZ lines splitting with simulations at the same time.

Despite the high strain sensitivity of CBED due to the HOLZ lines position, as well as their rocking curves, strain measurement using CBED configuration is not popular. Modern electron diffraction based methods mainly used nanobeam diffraction configuration (NBD) (Béché et al., 2009). In this mode, the sample is illuminated by a small parallel beam. The sample is then located in the pupil plane of the illumination system. ZOLZ spots are observed in the BFP of the objective lens, and their positions can be precisely fitted to determine the evolution of the lattice parameters if the NBD spot is scanned over the sample surface. These methods are indeed much easier to implement than CBED requiring dynamical simulations and finally containing too much informations for classical strain mapping applications. Nevertheless, recent developments of pixellated detectors for 4D-STEM applications where full CBED patterns could be acquired in STEM with high scan rate (around 100 ns per pixel on a 1024×1024 field of view), are opening new interesting perspectives for CBED based methods in strain measurement but more generally in quantitative field mapping, crystal symmetry determination, ... (Gammer et al., 2018).

Despite pushing these diffraction based directions, I chose instead to work on developments of new electron holography methods for application in strain measurement. The original idea was to develop easier procedures

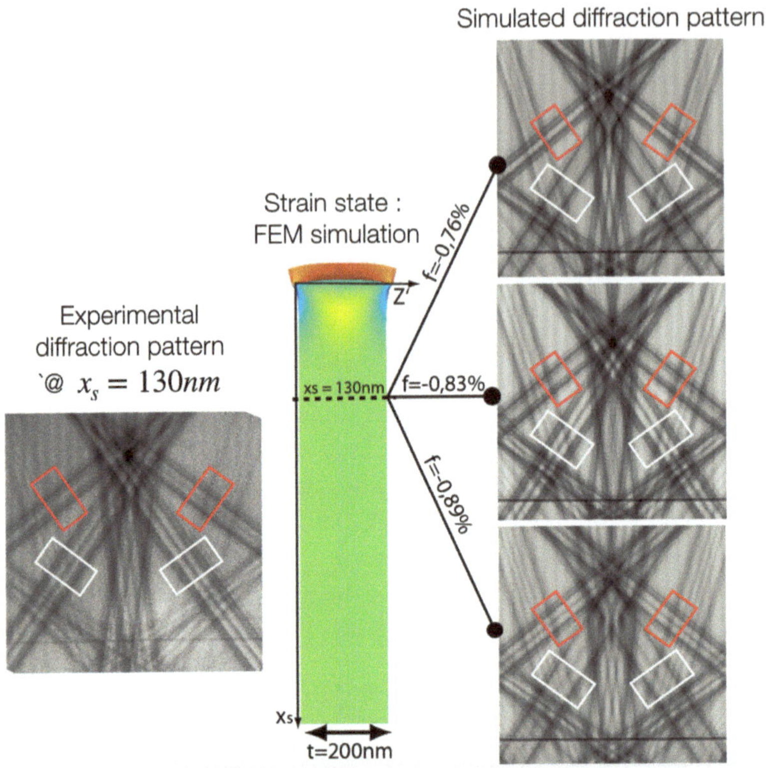

Fig. 1.15 Description of the final strain measurement method using CBED for single epitaxial layer: experimental HOLZ line profiles taken at a precise position inside the substrate must be compared to dynamical simulated ones obtained considering the thin lamella strain relaxation thanks to FEM simulations. *Extracted from Houdellier, F., 2006. Contribution au développement du CBED et de l'holographie HREM pour l'analyse des déformations de couches épitaxiées. PhD thesis INSA Toulouse and CNRS-CEMES (2006). INSA, Toulouse, PhD thesis.*

than CBED to be implemented on real devices without losing sensitivity or field of view. DFEH introduced in the next part is the first step toward these developments.

1.2.2 Strain mapping with large field of view using DFEH

1.2.2.1 Introduction to DFEH

Off-axis electron holography, introduced in Section 1.1.2.3, allowed to quantitatively map magnetic and electric fields at the nanoscale (Tonomura, 2011). However, electron holography is not limited to the measurement of magnetic and electric contributions on the electron phase shift. We have showed that a

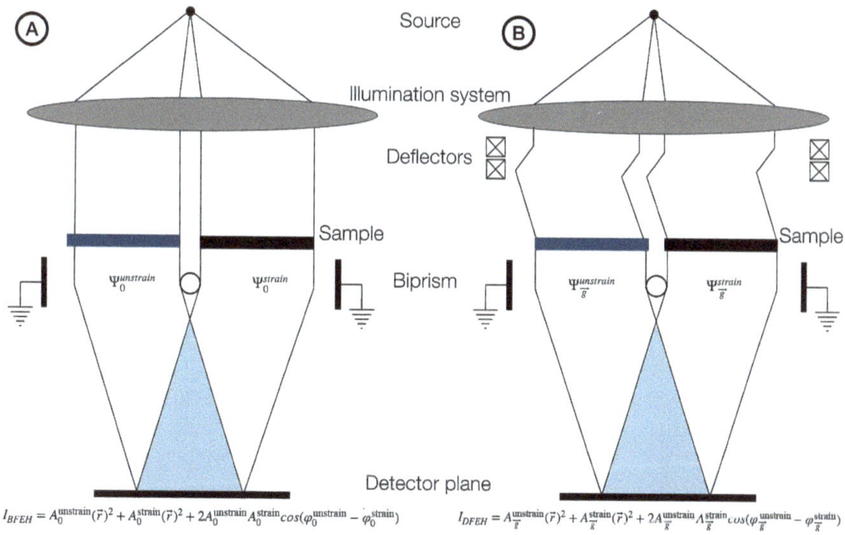

Fig. 1.16 Off-axis electron holography schemes: (A) Bright-field electron holography (BFEH) with strained crystal transmitted wave Ψ_0^{strain} and the unstrained one $\Psi_0^{unstrain}$. (B) Dark-field electron holography (DFEH) with strained crystal diffracted wave $\Psi_{\vec{g}}^{strain}$ and the unstrained one $\Psi_{\vec{g}}^{unstrain}$.

geometric electron phase shift induced by a displacement field $\vec{u}(\vec{r})$ located in a crystalline sample probed by the electrons beam, can also be measured and quantified directly by electron holography using the DFEH configuration described in Fig. 1.16B, and first introduced by Hanszen (Hanszen, 1986; Hÿtch et al., 2008a). The technique has opened up a new range of applications for measuring strain in crystalline materials, notably in the field of semiconductor devices and thin films (Hÿtch et al., 2011).

Geometric phase had previously been measured by HREM using Fourier image treatment methods (Hÿtch et al., 1998). But, as already mentioned for CBED, the effect of a displacement field $\vec{u}(\vec{r})$ on the HOLZ line rocking curve can be kinematically understood by adding a phase shift $\exp(-2\pi i\vec{g}.\vec{u}(\vec{r}))$ on the diffracted plane wave. So it is natural to think that geometric phase $\varphi_{geo} = -2\pi\vec{g}.\vec{u}(\vec{r})$ could be measured directly using interferometric methods such as off-axis electron holography.

The geometric phase is only present in the diffracted beams and therefore cannot be measured by the conventional off-axis electron holography setup where the interference is obtained using the transmitted beam as described in Fig. 1.16A. Standard off-axis holography is performed considering the reference beam taken in a vacuum area, i.e., $\Psi_0^{\text{unstrain}} = \Psi_0^{\text{vacuum}}$ which is then easier to analyze. In DFEH the crystal is oriented in diffraction conditions for one of the lattice planes $\vec{g}$ and the illumination conditions tilted so that the emerging diffracted beam is aligned with the optic axis, exactly as for conventional dark-field imaging, and an objective aperture is applied to eliminate the other diffracted beams (Hÿtch et al., 2011). A hologram is created from the interference between the diffracted beam emanating from an unstrained region of crystal $\Psi_{\vec{g}}^{\text{unstrain}}$, which serves as the reference, and a beam from the region of interest in the strained crystal $\Psi_{\vec{g}}^{\text{strain}}$ (see Fig. 1.16B). The relative phase difference of the diffracted beams between the two regions can then be determined from the dark-field hologram, assuming a suitable calibration of the carrier frequency, i.e., the tilt angle α generating by the electron biprism (see Fig. 1.5).

To measure the geometric phase component φ_{geo}, the other phase terms, notably crystalline and electrostatic phase assuming a nonmagnetic sample, must be eliminated. To a first approximation, these two phases do not depend directly on the local strain. Therefore, if the sample is uniformly thick, these terms will cancel out and a direct measurement of geometric phase difference between $\Psi_{\vec{g}}^{\text{unstrain}}$ and $\Psi_{\vec{g}}^{\text{strain}}$ becomes possible: $\Delta\varphi = \varphi_{\vec{g}}^{\text{unstrain}} - \varphi_{\vec{g}}^{\text{strain}} = \varphi_{\text{geo}}^{\text{unstrain}} - \varphi_{\text{geo}}^{\text{strain}}$.

Indeed, for the measurement of strain, which depends on the phase gradient, it is only necessary that the gradient of the nongeometric phase terms be zero, or the gradient of the difference. Fortunately, the requirements for uniform thickness are not as stringent as for the measurement of dopant distributions since the geometric phase variations are generally quite large, as the examples will show. In order to have a constant crystalline phase, the diffraction conditions must also be uniform over the two regions. The situation is more complex in the presence of compositional variations, as these will introduce additional crystalline and electrostatic phase terms. Significant and localized phase changes can occur at interfaces, which are therefore difficult to eliminate from the analysis. Finally, the measurement will provide the relative distortion of the reference lattice and the region of interest. Therefore, a strict mechanical strain analysis requires the reference region to be unstrained, otherwise a systematic error will arise. Note, that there is no fundamental reason why the reference region must be the same material as the region of interest, provided that it diffracts within the objective aperture. Only the interpretation requires a suitable adaptation. We will describe in Section 1.3.2 the specific configuration where a tilted reference wave

(TRW) in the vacuum can be used as a perfect unstrained reference (Röder et al., 2016). This original DFEH configuration was only possible thanks to the unique configuration of the I2TEM microscope.

We will call the equivalent setup for the transmitted beam, the bright-field electron hologram (BFEH). Here, the transmitted beam from the region of interest is interfered with that emanating from the reference region of crystal, and not with the beam in the vacuum as for the conventional setup (see Fig. 1.16A). This configuration is of primary interest for assessing, and correcting the systematic errors of the dark-field measurements since for the transmitted beam, the geometric phase component should be zero (Hÿtch et al., 2011).

1.2.2.2 DFEH with large field of view thanks to Lorentz mode

The configuration of DFEH reported in Fig. 1.16 has been simplified on purpose compare to its real practical implementation in a TEM. Indeed, the hologram is never directly generated from the object wave, i.e., with a biprism physically located in an optical plane below the object, but using the image wave formed after the objective lens. From a paraxial point of view, either in standard off-axis holography, BFEH, or DFEH, the phase difference $\Delta\varphi$ will be strictly identical with the one obtained without considering the objective lens effect. Indeed under paraxial approximation the objective lens will not affect $\Delta\varphi$, but only change the hologram field of view relatively to the image size. Usually, an electron biprism will be installed between the objective lens yoke and the first intermediate lens very close to the selected area aperture plane where the object magnified image is obtained (Völkl et al., 1999). The off axis hologram will then be recorded in this image plane and the field of view W, compared to the image dimension, will then depend on the objective lens magnification ($(x|x)$ in Eq. 1.2). It is worth noted that aberrations have to be considered in high resolution holography configuration (Völkl et al., 1999). In this specific off-axis configuration the sample is oriented in ZAP (such as in the case of HREM imaging) and, if the magnification of the objective lens is sufficient, it becomes possible to separate in the hologram FFT, the interference coming from the transmitted beam and the vacuum reference area, with all the other interferences generated by the superposition of each diffracted wave and the same vacuum reference beam. In that particular case the objective lens will then affect the quantities $\Delta\varphi_{\vec{g}/0}$, measured from the interference of a diffracted beam $\vec{g}$ and the vacuum reference beam 0, depending on the selected $\vec{g}$. That's why, high resolution electron holography is usually used to retrieve the main objective lens wavefront aberrations, such as the aperture one, in order to correct them using specific postprocessing methods applied to the hologram sideband (Orchowski et al., 1995).

As far as DFEH is concerned, we will only consider the paraxial effect of the objective lens as the two diffracted wavefronts (the "unstrain" and the "strain") will be affected by the same aperture aberrations which are canceled out in the measurement of $\Delta\varphi = \varphi_{geo}^{unstrain} - \varphi_{geo}^{strain}$. To be more specific, only objective lens off-axial aberrations, which depend on the object spatial coordinate x_0, y_0, such as coma $(x|xaa)$ depending linearly on x_0 (together with equivalent term along y, and mixed complex terms), could affect the measurement. However it is reasonable to neglect them as far as the hologram field of view remains in the range of hundred nanometer or below, remaining in isoplanatic condition (Welford, 1976).

Carrying out an holography experiments to map the magnetic configuration of a sample in a TEM is impossible using standard objective lens. The sample is indeed located in the center of the objective lens pole piece gap where a field in the range of few Tesla is applied along the optic axis, i.e., perpendicular to the sample (Völkl et al., 1999). This field will saturate the magnetic configuration of the object under analysis. The operator will then have to stop the objective lens field. This is known as "low magnification" configuration. Without objective lens, the object will not be sufficiently magnified inside the hologram width W. The answer is to use a Lorentz lens. Most of the time, this lens is optically located below the objective-image lens and the focal length f_{Lor} of this lens is then bigger that the standard objective lens $f_{obj} < f_{Lor}$. The use of this lens will have two important consequences on holography experiments. First of all the object could then be magnified inside the hologram without the use of the main objective lens, leaving the sample in a magnetic free environment. But, most importantly regarding the DFEH configuration, and as reported in Fig. 1.17, the use of a Lorentz lens will decrease the image size inside the hologram width W. This second consequence, which could be problematic to study small magnetic objects, is however very positive for DFEH analysis. Indeed, main applications of strain mapping are found in the microelectronic community. The size of their stressed devices as well as the localization of an unstrained zone relatively to the strained area, prevent the use of DFEH configuration using standard objective lens due to its strong magnification. This limitation of DFEH can be relaxed using the Lorentz lens configuration where few hundred of nanometers field of view DFEH can be recorded, which has proven to be largely sufficient to study modern devices, as we will see in the following (Hÿtch et al., 2008a, 2011).

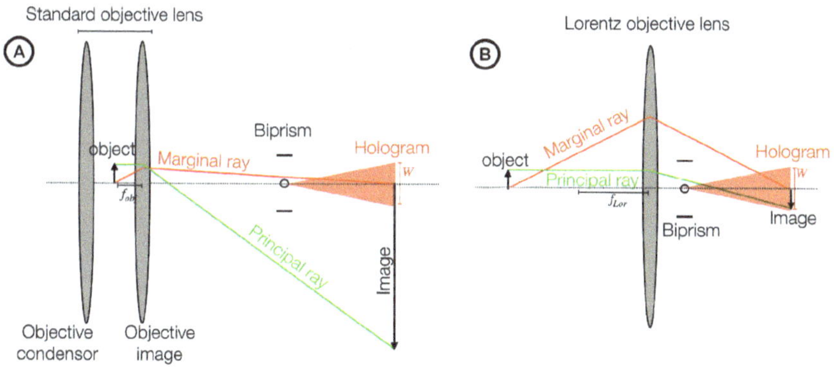

Fig. 1.17 Paraxial fundamental rays showing the influence of the objective lens magnification on the hologram field of view. (A) Standard TEM configuration. (B) Lorentz lens configuration.

1.2.2.3 Strain measurement using DFEH

The test sample used to illustrate the technique is an array of dummy p-MOSFET transistors with embedded $Si_{0.7}Ge_{0.3}$ sources and drains (Hÿtch et al., 2008a). TEM specimens were prepared by focused ion beam (FIB) to a thickness of between 100 and 150 nm. Observations were carried out on the SACTEM-Toulouse (Houdellier et al., 2008b). Specimens were oriented close to a [−110] zone axis. The microscope is operated in the Lorentz mode using the corrector first transfer lens as a Lorentz lens and with the main objective lens and hexapoles switched off (see Fig. 1.21). Typical fringes spacing are $s = 1$–2 nm, fringes contrast of around $C = 20\%$, and hologram widths from $W = 300$–500 nm allowing lengthwise fields of view of several microns. Strain fields are extracted using homemade software coded in Gatan Digital micrograph language (Hÿtch et al., 2008b).

Once the hologram is recorded the geometric phase can be retrieved. The process used is based on the Fourier transform described by Eq. (1.19) and mapped on Fig. 1.18.

It requires the correct choice of side band, which can be understood using Fig. 1.18 (Hÿtch et al., 2011). The crystal diffracts into the transmitted beam O and diffracted beams from the undistorted crystal, A and A', and the strained region of crystal, B. The objective aperture is used to select only these diffracted beams. The biprism deflects these beams depending on whether they pass to the left (A) or the right (A', B). The carrier frequency linked to the biprism deflection angle α is denoted by q_c. In this example, the voltage of the biprism is positive and the wire is then located before the image plane. As usual, the Fourier transform of the holographic fringes will

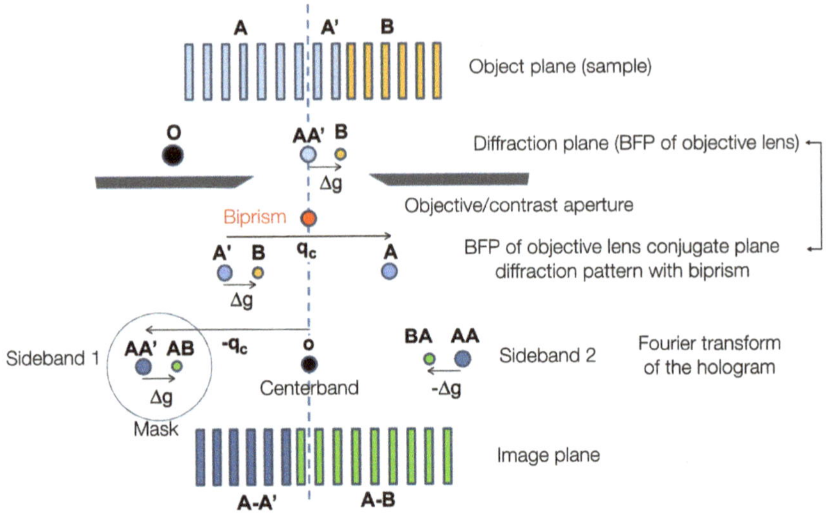

Fig. 1.18 Side-band selection for dark-field holography. See text for details. *Extracted from Hÿtch, M.J., Houdellier, F., Hüe, F., Snoeck, E., 2011. Dark-field electron holography for the measurement of geometric phase. Ultramicroscopy 111 (8), 1328–1337.*

have two side bands (see Eq. 1.19) (Völkl et al., 1999). It can be seen that the direction of $\Delta\vec{g}$, corresponding to the difference in reciprocal lattice vectors between strained and unstrained crystal areas, is unaltered in the side band pointing toward the substrate. It is this side band that we choose to reconstruct. Using the other side band simply changes the sign (Hÿtch et al., 2011).

The reference region of phase A–A' can be conveniently used to define both the unstrained state and the carrier frequency. Any phase gradient elsewhere in the hologram indicate a difference in local lattice parameter or orientation due to the quantity $\Delta\varphi = \varphi_{\text{geo}}^{A-A'} - \varphi_{\text{geo}}^{B}$. The strain in the direction of the diffracting vector $\vec{g}$, can be determined directly from the gradient of the phase $\Delta\varphi$. To determine $\vec{g}$ in image coordinates, the corresponding lattice spacing $d_{\vec{g}}$, needs to be known along with the magnification of the hologram and the direction of the diffraction vector with respect to the image. Indeed, the difference in reciprocal lattice vector $\Delta\vec{g}$, can be determined unambiguously from the hologram but to determine percentage changes in lattice parameters and orientation the length and direction of $\vec{g}$ is required. For instance, using the $\vec{g} = (220)$ diffracted beam in the (001) grown strained-silicon transistor sample, the strain could be measured in the direction parallel to the specimen surface from source to drain, corresponding to the ε_{xx} component in Fig. 1.19.

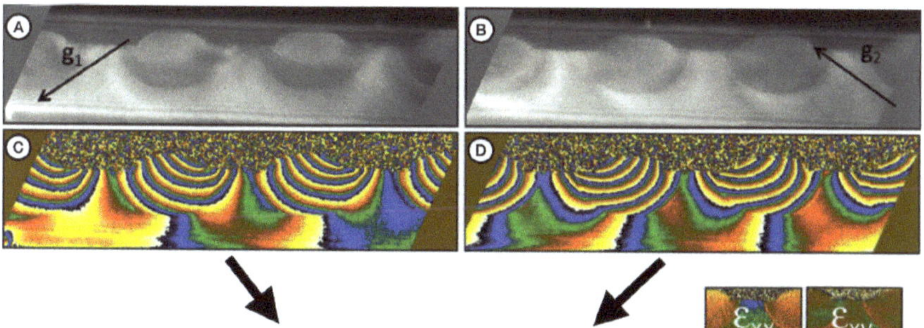

$$\vec{u}(\vec{r}) = -\frac{1}{2\pi}\left(\Delta\varphi_{geo}^{\vec{g}_1}.\vec{a}_1 + \Delta\varphi_{geo}^{\vec{g}_2}.\vec{a}_2\right)$$

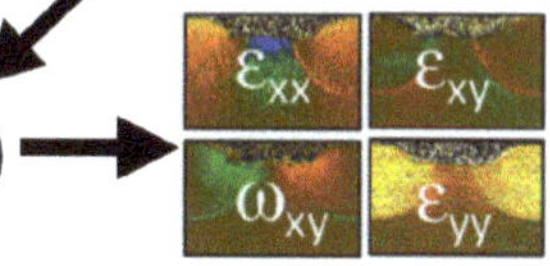

Fig. 1.19 Determination of two-dimensional displacement field by dark-field holography: (A) dark-field hologram for (−1 − 1 − 1) beam; (B) dark-field hologram for (−1 − 11) beam; (C) extracted phase of (−1 − 1 − 1) DFEH; (D) extracted phase of (−1 − 11) DFEH. 2D displacement calculation and corresponding strain tensor components are also reported. *Extracted from Hÿtch, M.J., Houdellier, F., Hüe, F., Snoeck, E., 2011. Dark-field electron holography for the measurement of geometric phase. Ultramicroscopy 111 (8), 1328–1337.*

To determine the 2D strain tensor, holograms need to be taken for two noncollinear diffraction vectors $\vec{g}_1$ and $\vec{g}_2$ (Hÿtch et al., 2011). For the [110] zone-axis sample, we typically chose two (111) diffracted beams. The result obtained from these experiments is shown in Fig. 1.19. They can finally be combined to determine the full in-plane strain tensor ε_{ij}, and rigid-body rotation ω_{ij}. The difficulty is that the holograms need to be aligned correctly but fortunately, this alignment procedure is facilitated in that the technique is a medium-resolution technique. We either use the phase images or the amplitude images to identify equivalent sample areas.

Such as for CBED measurements, DFEH evaluations have to be compared with FEM simulations of the sample strain state in order to take into account thin film relaxations. Fig. 1.20 reports the comparison for the p-MOSFET transistors used to develop the DFEH method. We can notice that the overall agreement between the experimentally determined strain maps and those coming from the modeling is excellent (Hÿtch et al., 2008a, 2011).

1.2.2.4 Summary on strain mapping using DFEH

By implementing dark field illumination inside an FE-TEM, we show that geometrical phase $\varphi_{\text{geo}} = 2\pi\vec{g}\vec{u}(\vec{r})$ is a real physical quantity that could be extracted directly from the diffracted beam phase thanks to interferometric method. Carrying out an interference between two $\vec{g}$ reflections taken from an unstrained and a strained area, it was possible to measure directly the

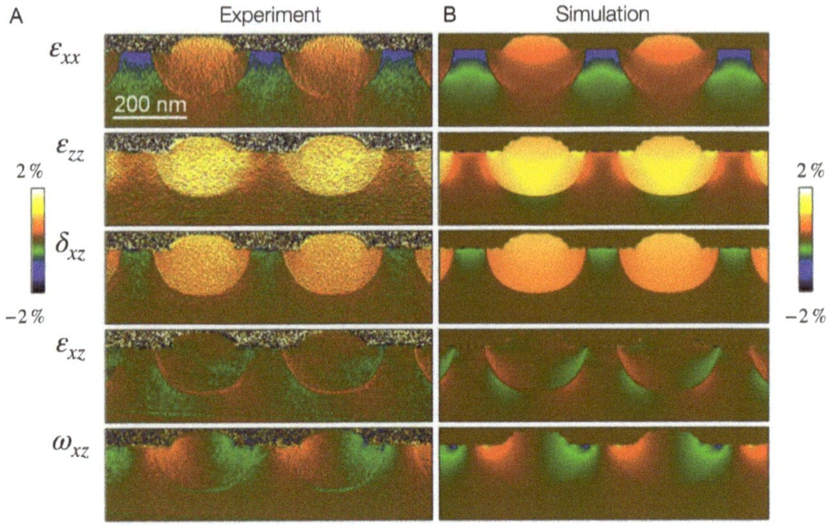

Fig. 1.20 Strain tensor for p-MOSFET transistors: (A) experimental strain components and (B) simulated strain components from finite element method. *Extracted from Hÿtch, M., Houdellier, F., Hüe, F., Snoeck, E., 2008. Nanoscale holographic interferometry for strain measurements in electronic devices. Nature 453 (7198), 1086–1089.*

geometrical phase difference $\Delta\varphi$ and thus extract the associated displacement field $\vec{u}(\vec{r})$. DFEH is then a fully quantitative technique for the measurement of strain in nanostructures and devices. The analysis is straightforward and gives directly the strain components that can be compared with modeling. DFEH is, in particular, adapted to epitaxial systems as the substrate can act as a reference area. The technique has already been applied successfully to a number of systems, from the MOSFET and FinFET devices and similar strained silicon devices, to strained layers, misfit dislocations therein and quantum dots (Boureau et al., 2020; Cherkashin et al., 2013; Hÿtch et al., 2011). The technique can be powerfully combined with conventional holography to provide a complete study of strain and dopants in devices. Finally, the implementation of this configuration using a Lorentz objective lens allowed us to map this displacement field over wide field of view (almost 1 μm).

However, the main difficulty of DFEH was related to the Lorentz configuration used in the SACTEM-Toulouse. Our main limitation lied in the accessible illumination tilt angle in dark field before the beam touched mechanically the upper part of the objective lens which remains inactive. As a consequence, we were limited to low $\vec{g}$ reflections to record DFEH (maximum $\vec{g} = (004)$ in the silicon at 200 kV). Measuring very small displacement field $\vec{u}(\vec{r})$ will benefit in selecting higher order of $\vec{g}$ reflections as the geometrical phase sensitivity increases with $\vec{g}$. Our goal was to develop a new instrument allowing to reach higher reflections such as the $\vec{g} = (008)$ of the silicon, and enable the implementation of modern correction of biprism artefacts, described in Section 1.1.3.3, using multibiprism optics (Harada et al., 2004). Furthermore, this new instrument will be designed to allow split beam illumination using a biprism located in the illumination system (Tanigaki et al., 2012). Split beam can be used to select the reference area independently from the measurement area in off-axis electron holography experiments. This configuration could be very interesting either to map electric or magnetic fields using BFEH as well as strain fields using DFEH. The development of this instrument, called the I2TEM, started around 2009. It was performed in collaboration with the Hitachi company and will be the topic of the next part.

1.3. Improving coherent TEM

1.3.1 Introduction

Carrying out coherent TEM with enough signal to noise ratio in the detected signal requires brightest electron source, ideally a Dirac-like point source. In this latter ideal situation, all electrons will come from the same spatial origin and each plane wavefront observed far away from the source will then have a phase relation over the two infinite dimensions perpendicular to the propagation direction. In that case, the coherent current available to perform an electron hologram, given by Eq. (1.37), will then be maximum. Of course, even the brightest electron source available nowadays (but honestly designating with confidence one specific technology among existing developments is not very wise ...) will have an extended spatial distribution at the origin. As described in Section 1.1.3.2, for a given spatial distribution $s(\nu_q)$, the dimensions of the plane wavefront, observed far away from the source, on which a phase relation remains will be reduced compare to the ideal point source. Increasing the source intensity distribution $s(\nu_q)$ will, of course, decrease the lateral dimensions of the coherent wavefront (Born and Wolf, 1999; Gross, 2005).

There are two possible directions to reduce the size of $s(\nu_q)$ to improve coherent TEM methods that we have investigated in all our different developments. But first, it is worth noting that the relevant distribution which have to be considered for a coherent experiments will be the last one before the object. This crossover will be formed by the illuminating optics located between the source and the object. The optical configuration chosen to illuminate the sample will then have a very important impact on the experiment. Therefore, the first simple way to reduce the object illumination crossover will be to strongly demagnify the original source crossover. But, due to illuminating optics aberrations, especially the aperture one, the use of illumination apertures will then become necessary to enable the formation of small crossover. This will automatically result in a loss of probe current as the source brightness is conserved from the source, to the detector (Born and Wolf, 1999). As a consequence, higher demagnification will result in higher loss of currents, and this method cannot be used if the original source brightness is not sufficient, i.e., using thermionic source where the need of very strong demagnification will result in dramatic loss of currents.

However, even with high brightness source, the illumination optics have to be optimized to generate a suitable crossover depending on the experiments (STEM, holography, ...). This optimization is particularly important in off-axis electron holography where the use of a too high demagnification, and the associate probe current loss, will result in the need of higher exposure time preventing the observation of contrasted fringes due to the contribution of instabilities. However, hologram fringes recorded under off-axis configuration are generated by the interference of the two half-wavefronts located on each side of the biprism wire. Optimizing the fringes contrast will require a higher demagnification only along that direction. That's the reason why elliptical illumination, with an ellipse direction perpendicular to the biprism wire, are usually implemented in off-axis holography experiments (Lehmann, 2004). Using such kind of illumination, the source crossover before the object will therefore have an elliptical shape, with a very small dimension perpendicular to the wire, maximizing the degree of coherence along that direction, and a biggest dimension along the wire to optimize the hologram total current.

Optimizing coherent TEM experiments using dedicated and optimized column optic, knowing that the commercial source defined by its brightness cannot be changed, corresponds to the direction taken for the development of the I2TEM column I want to describe in the following. The second direction will be of course to address the problem directly at its source, optimizing the brightness of the brightest technology available, i.e., the CFE source (Hawkes, 2009). The developments performed following that direction will be described in Chapter 2 dedicated to instrumental developments.

1.3.2 Improving the optical flexibility of electron interferometry: The development of the I2TEM

1.3.2.1 Target specifications

Before the installation of the I2TEM microscope, electron holography in CEMES, and particularly DFEH experiments, were performed using the SACTEM. Hence, before describing the unique characteristics of the I2TEM instrument, I would like first to present the configuration of the SACTEM to stress out its limitation regarding holography experiments. As quickly introduced previously, the SACTEM is a TECNAI F20 based microscope from the FEI company, i.e., a microscope equipped with a 200 keV Schottky source, two condenser lenses, a super twin objective lens and finally two intermediate and two projector lenses. This microscope was then retrofitted with a semiaplanetic system C-COR from the CEOS GmbH company located after the objective lens and one rotative electron biprism located in the SA aperture position, limiting the number of accessible SA apertures to two instead of four ordinary. The main limitations of this configuration for DFEH arise from the position of the single biprism and the configuration of it's Lorentz optical mode. They can be briefly summarized below:

1. The position of the electron biprism in the same plane as the SA aperture prevents the formation of an electron hologram in the SA mode of the TEM. Yet, SA mode includes almost all the intermediate magnifications of the imaging lenses. Acquiring an hologram using these intermediate magnification required the use of free lens control and suffered from a strong lack of optical flexibility. Additionally to a poor optical flexibility we must also include the annoying interconnexion between the hologram width W with the interference fringes distance s. Indeed, such as for all single biprism off-axis configuration, increasing W will increase the deflection angle α generating hologram with smaller s. Adding all these limitations together, it becomes very challenging to prepare a random experiment with a required field of view, set by W and the magnification, independently from the hologram resolution, defined by s.
2. Carrying out Lorentz imaging using the SACTEM required a specific setting of the C-COR reported in Fig. 1.21. Fig. 1.21A shows the paraxial trajectories inside the C-COR when all the optical elements are activated (round transfer lenses (TL), adaptive lens (ADL) as well as the two long hexapoles (Hex)) (Rose, 1990). When the main objective lens is OFF, Lorentz imaging condition can be implemented by shutting down all the C-COR lenses but the TL_{11}. The two paraxial trajectories of this

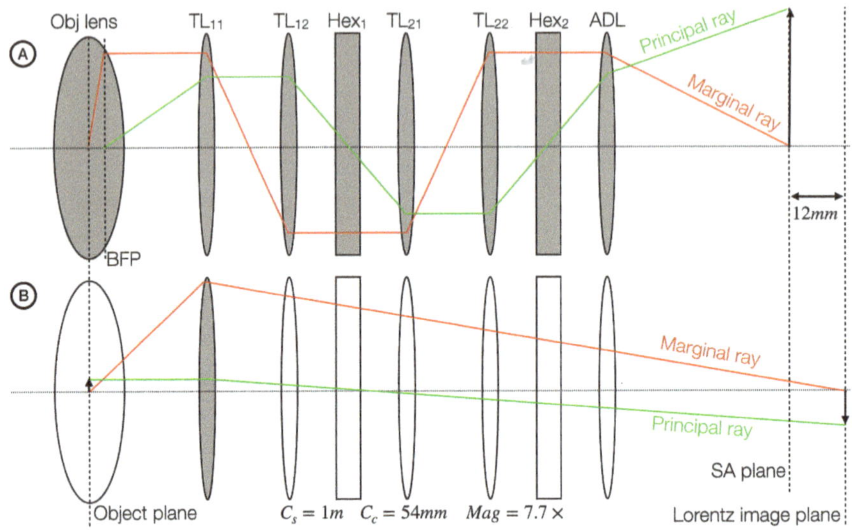

Fig. 1.21 Paraxial fundamental rays (principal and marginal rays) inside the SACTEM C-COR system. (A) Standard corrected mode. (B) Lorentz TL_{11} mode.

so-called TL_{11} Lorentz mode are reported in Fig. 1.21B (Snoeck et al., 2006). On purpose, the image plane is slightly shifted below the standard SA plane due to the previous point related to the biprism location. Recording hologram in Lorentz mode was then possible using the SACTEM thanks to this new condition. As a result of the mechanical dimensions, a linear magnification of 7.7× could be reached. This value was sufficient for most of experiments, but the global image quality was very poor mainly due to the strong aperture aberration ($(x|aaa) = (y|bbb) = (r|aaa) = C_s = 1\,m$). Moreover, since the beam deflections coils used to tilt the beam on the sample plane are located before the objective lens pole piece and that the objective lens remains OFF in the Lorentz mode, the achievable tilts were limited to 15 mrad. Above 15 mrad the incoming beam touches the upper pole piece hole. Recording DFEH also requires the use of contrast aperture located in the BFP of the objective lens to select the proper $\vec{g}$ reflection. Unfortunately, using TL_{11} Lorentz mode, the BFP of the TL_{11}, located inside the C-COR, was not mechanically accessible (see Fig. 1.21B). The use of a contrast aperture located in an optical plane different from the optimum aperture-stop position strongly affects the imaging field of view of the DFEH imaging mode. Indeed, as observed in Fig. 1.21B, the aperture mechanically located in the objective lens BFP will cut the principal ray when TL_{11} Lorentz mode is used. These two last points were the main limitations for DFEH applications.

3. Finally, the pole piece gap of the super TWIN objective lens was 4.5 mm. This last point was also a limitation regarding all the *in situ* applications targeted in the laboratory such as contacting devices to study transport properties of devices, applying external magnetic fields on the sample, ... while recording off-axis hologram at the same time. These specific experiments will indeed require the use of bigger holder maintaining standard tilting capabilities around 40 degrees, or more, which will not be possible with the gap of the SACTEM pole piece.

Resolving all these limitations and propose a dedicated electron interferometry platform were the goals behind the developments of the I2TEM. The main specifications required can be briefly summarized below:

1. A versatile optic designed for optimum off-axis holography where the magnification of the sample inside the hologram, the field of view W as well as the resolution s can be set independently. This will require the use of multibiprism configuration in the imaging part, and at least one biprism carefully located in the illumination optic (Harada et al., 2004, 2006; Tanigaki et al., 2012). Mechanically all the biprism holders will have to be located in dedicated positions separated from standard aperture positions.
2. The microscope should offer an optimized Lorentz optical mode allowing to perform aberration corrected Lorentz imaging with the same flexibility as standard TEM imaging. Tilting the beam should be possible to almost 30 mrad to perform DFEH with (008) reflection of Silicon. Mechanically, the contrast aperture should be also located in the correct aperture-stop position of the Lorentz mode to avoid any vignetting from the aperture in the recorded DFEH. While image field of view obtained with Lorentz mode are bigger than using standard TEM objective lens, the aberration correction should be aplanatic to minimize contribution of off-axial aberrations such as coma (Welford, 1976). Applying an external magnetic field on the sample using a controlled excitation of the main objective lens current while observing the sample in Lorentz mode should be also possible.
3. Finally, the mechanical configuration of the sample area must be carefully designed allowing the use of bigger *in situ* sample holders with adapted sample tilt range.

The final target of this instrument was to optimize standard off-axis electron holography technique as well as develop new electron interferometry configurations and perform *in situ* observations of sample using them.

1.3.2.2 I2TEM column overview

Fulfilling the specifications described previously, have requested the development of a specific instrument not available originally on the market. To perform this challenging development we jointly worked with the Japanese manufacturer Hitachi-High Technologies (HHT) Corporation located in Hitachinaka in the north of Tokyo area (Iida and Taniguchi, 2016). The I2TEM project was financially supported by a "CPER" (Contrat Plan Etat Région) and the "equipex MIMETIS." I2TEM was the beginning of our joint adventure with HHT which have grown in intensity after the installation of the I2TEM, especially around the developments of new CFE technologies that I will describe in detail during Chapter 2.

Together with HHT and CEOS GmbH located in Heidelberg Germany, we worked on the instrument described in Fig. 1.22. The overview of the optical configuration is reported in Fig. 1.22A and was adapted from a standard HF3300 microscope. The illumination system is composed by a 300 keV CFEG source, followed by a first condenser aperture (cond apt), a rotatable electron biprism (*BPC*) and three condenser lenses (C_1, C_2, C_3). A second STEM aperture (STEM apt) is located between C_2 and C_3. The first condenser aperture is used in the TEM illumination mode described in Fig. 1.2A while the STEM aperture is used in the STEM illumination mode schematically reported in Fig. 1.2B.

The objective lens has been selected with a wide pole piece gap of 10 mm following the *in situ* requests previously described. Two sample stages can be inserted in the objective area. The first one is located above the objective lens yoke, 11 mm below the lower pole piece of the C_3 probe forming lens and is called Lorentz stage. The second is located in the standard position, allowing to insert the sample holder directly in the middle of the objective lens pole piece gap. This stage is called normal stage.

The corrector is an aplanatic solution originally developed by the CEOS company and called B COR (Müller et al., 2011). As briefly described in the introduction and reported in Fig. 1.3D, the CEOS B COR is a $8-f$ system based on an hexapole triplet. Two transfer lenses doublets (respectively called TL_{21}/TL_{22} and TL_{31}/TL_{32}) are placed between them. The first transfer doublet is located between the first and second hexapole planes (called Hex_1 and Hex_3). The second one is installed between the second and third hexapole planes (Hex_3 and Hex_5) (Müller et al., 2010).

The hexapole fields successively contribute to the aperture aberration correction such as in standard C COR. For the correction of off-axial coma two additional weak hexapole doublets (respectively called Hex_{21}/Hex_{22}

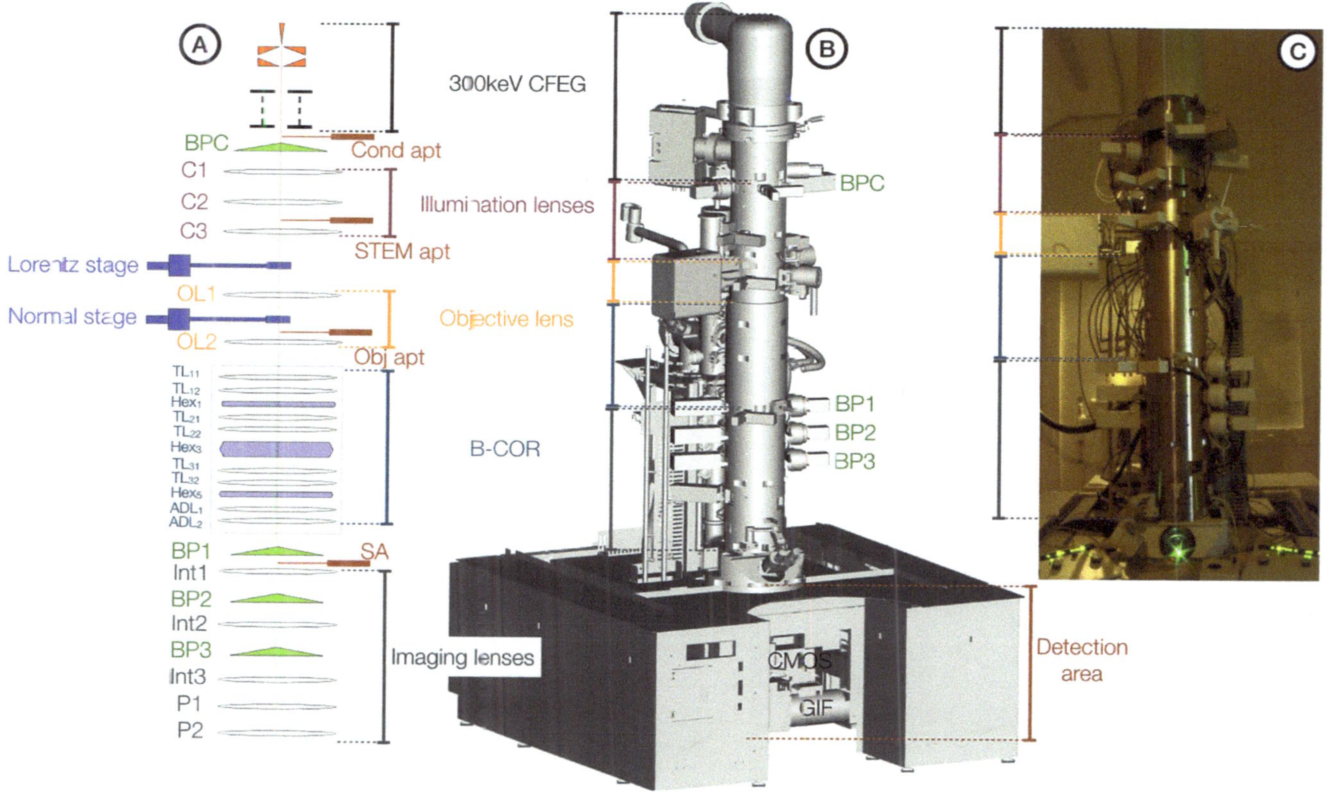

Fig. 1.22 Outline of the I2TEM instrument. (A) Optical description of all the optical elements. (B and C) External view of the instrument (CAD and real instrument). All the elements described in (A) are reported using their specific color.

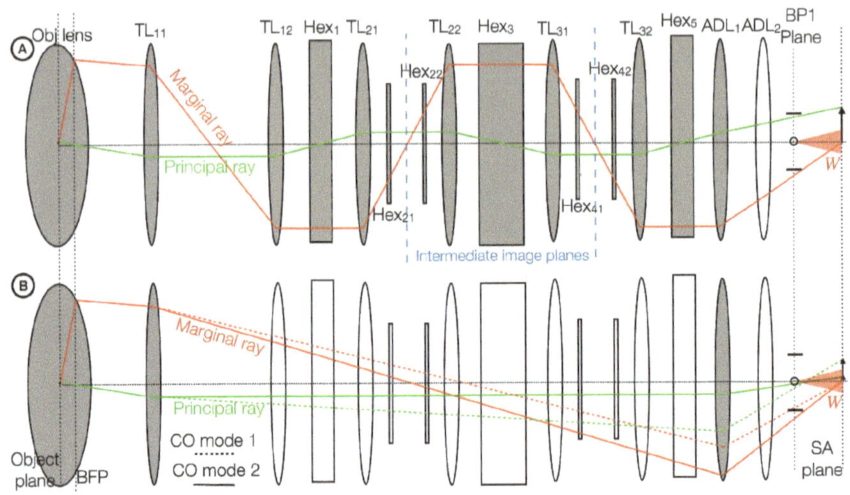

Fig. 1.23 Paraxial fundamental rays (principal and marginal rays) inside the I2TEM B COR system. (A) Standard aplanatic corrected mode. (B) Lorentz TL_{11} mode. Two cross-over (CO) modes are represented showing the influence of the B COR optic on the hologram field of view bringing higher flexibility compare to standard TL_{11} mode of a C COR.

and Hex_{41}/Hex_{42}), not represented in Fig. 1.3D but reported in Fig. 1.23A, are excited antisymmetrically relatively to the intermediate image planes. These two intermediate images planes are located in the middle of the first and the second transfer lens doublets, i.e., between TL_{21} and TL_{22} then between TL_{31} and TL_{32} as reported in Fig. 1.23A.

All second-order aberrations, locally introduced by the hexapole fields, vanish outside the corrector. Fig. 1.23A reports the two paraxial trajectories when the corrector is fully activated. The corrector is coupled to the objective lens by means of a further transfer lenses doublet (TL_{11}/TL_{12}) and a final adapter lenses doublet (ADL_1 and ADL_2) is used to form a magnified image at the selected-area (SA) plane below the corrector and before the first intermediate lens (I_1). The necessary column extension for the standard Hitachi HF3300 amounts to 430 mm and the contribution of the corrector's focusing elements to the total chromatic aberration is about 25% of the chromatic aberration of the objective lens (Massmann et al., 2011). This last point is the only drawback which, together with the influence of the large pole piece gap, strongly limits the applications of I2TEM at low voltage.

Lorentz imaging using the B COR can be implemented similarly to C COR Lorentz mode optic used with the SACTEM, and described in Fig. 1.21. Main objective lens is switched OFF, and the TL_{11} lens is used as a Lorentz lens to focus an image in the SA plane. Compare to the

SACTEM configuration, the image plane of TL_{11} Lorentz mode of I2TEM remains in the standard SA plane as the electron biprism is inserted mechanically above this plane thanks to a dedicated port. Furthermore, aberrations of the I2TEM TL_{11} Lorentz imaging mode can be corrected as for the standard objective lens, maintaining aplanatic optical condition. Carrying out observations in Lorentz mode can then be realized over wide field of view.

The I2TEM B COR has also a specificity compared to the original design of CEOS. Indeed, the mechanical position of the TL_{11} lens relatively to the objective lens pole piece, has been optimized to improve the Lorentz imaging quality. *Lorentz type BCOR* distance between TL_{11} and the standard objective lens is 53 mm while this distance is 80 mm for the so-called *HIRES type B COR* originally proposed by CEOS (Massmann et al., 2011). This change has important consequences especially looking at the chromatic aberration of the Lorentz TL_{11} corrected mode going from $C_c = 520$ mm using the HIRES type B COR to $C_c = 164$ mm using the Lorentz type B COR. This improvement leads to an increase of the Lorentz imaging resolution going from 1 nm using HIRES type to 0.56 nm with the Lorentz type at 300 kV. Additionally, the image magnification on the SA plane is also affected by this height difference of the TL_{11} lens going from 4.2× using HIRES type to 8.8× with the Lorentz type. The I2TEM B COR is then optimized to perform Lorentz imaging sacrificing resolution in high resolution, when the main objective lens is activated. Indeed, a small increase of the chromatic aberration is observed in this mode going from $C_c = 3.5$ mm with the HIRES type B COR to $C_c = 3.7$ mm for the Lorentz type B COR. This will slightly degrade the standard resolution of the microscope going from theoretical resolution of 0.082 nm using the HIRES type to 0.084 nm using the I2TEM Lorentz type B COR at 300 kV. Working at this voltage, we have considered this difference insignificant. However, once again, this will have consequences for observations at low acceleration voltage for which I2TEM optics is not optimized. All these important parameters of the Lorentz type B COR TL_{11} mode, including aperture and chromatic aberrations coefficients, have been reported in Fig. 1.26.

Finally, compared to the C COR system, the B COR offers additional optical flexibility to adapt the image magnification in the SA plane. Since the first biprism (BP_1) is located before the SA plane, first transfer lenses of the corrector can be excited allowing to form a real crossover inside the corrector and before the adaptive lenses. Adjusting this crossover position will then adjust the image magnification inside the hologram width W. This adjustment is reported in Fig. 1.23B where two crossover modes (CO modes) obtained thanks to TL_{11} are shown. The optical flexibility given by the high number of transfer lenses located between the object and the first biprism of the B COR system, is then another interesting asset for off axis electron holography experiments (Linck, 2018).

As already described, the first rotatable biprism $BP1$ is inserted between the B COR system and the I_1 lens. Optically the wire is located before the SA plane. The imaging system of I2TEM consists in three intermediate lenses I_1, I_2, and I_3, and two projector lenses P_1 and P_2. Two additional rotatable electron biprism systems called $BP2$ and $BP3$ are inserted between I_1/I_2 and I_2/I_3 respectively. Biprism voltage can be adjusted between -500 and 500 V. Final acquisition is performed either using a Gatan one-view CMOS $4k \times 4k$ camera or using a standard CCD $2k \times 2k$ installed at the end of a post column Gatan Imaging Filter (GIF) quantum ER type.

I would like now to describe few specific features of the I2TEM which make this optical column unique for applications in electron interferometry.

1.3.2.3 Double stage Lorentz configuration

Thanks to the double stage configuration, the I2TEM is able to fulfill all the specifications required for the Lorentz imaging reported in Section 1.3.2.1. Carrying out Lorentz imaging is possible thanks to three distinct modes: single focus mode, double focus mode and TL_{11} mode. The last one has been described in detail in the previous section, and is used when the sample is inserted using the normal stage with the main objective lens OFF.

We will now describe the two other modes given once the sample is inserted using the Lorentz stage. In that case, the object plane is located almost 50 mm above the center of the main objective lens pole piece gap. The objective lens can first be used to focus the image on the standard SA plane enabling observation in Lorentz mode as the object is located in a magnetic field free area. Focusing using such high object distance requires a weak excitation of the objective lens. This mode, called *single focus mode* requires only few amperes in the objective lens to form an image in the SA plane. The two paraxial rays obtained in Lorentz single focus mode are reported in Fig. 1.24A. Compare to the TL_{11} Lorentz mode, the corrected single focus mode offers a slightly higher imaging resolution of 0.48 nm at 300 kV, due to a smaller contribution of chromatic aberration $C_c = 115$ mm, and a higher magnification of the image in the SA plane of 13×. More importantly, due to the mechanical position of the object, the single focus mode is no more limited in beam tilt range by the edge of

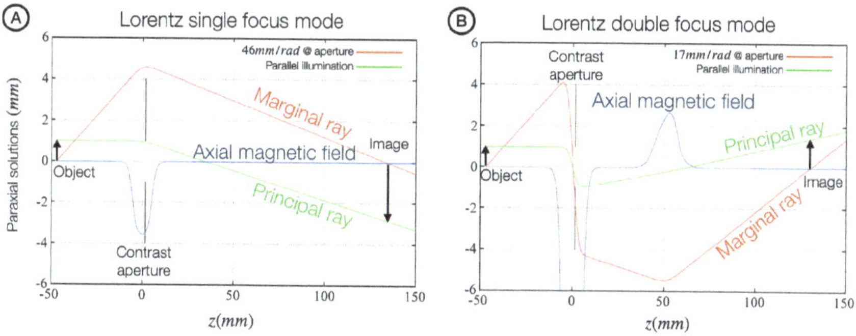

Fig. 1.24 Paraxial fundamental rays (principal and marginal rays) of the Lorentz imaging mode when the sample is inserted using the Lorentz stage. (A) Single focus Lorentz mode. (B) Double focus Lorentz mode.

the upper objective lens pole piece contrary to the TL_{11} Lorentz mode limited to 15 mrad tilt angle like the SACTEM.

However, knowing that the lower and upper dipoles positions of the beam deflector (BD) are located respectively 71 and 107 mm above the Lorentz stage object plane, the beam tilt remains nevertheless geometrically limited to 15 mrad by the maximum excitation offered by the upper coils, as described in Fig. 1.25A. This limitation is true only if the propagation remains field-free between the deflector and the object plane which corresponds to $C_3 = 0$A as the deflection coils are located between C_2 and C_3. However, the achievable beam tilt relatively to the Lorentz stage position can be increased by a factor 2 if the excitation of C_3 is set to 1.1 A, as reported in Fig. 1.25B. In that case dark field imaging with a beam tilt range of almost 30 mrad at 300 keV can be achieved when the sample is located in the Lorentz stage whatever the Lorentz mode, which was one of the required specification.

On the other hand, the single Lorentz mode cannot be used to perform a suitable dark field imaging condition with the contrast aperture properly located in the aperture stop position, as required initially in the specification. Indeed, as seen in Fig. 1.24A using single focus mode the principal ray doesn't cut the optic axis in the contrast aperture location (slightly below $z = 0$ mm), but somewhere close to $z = 40$ mm inside the B COR system where no aperture can be inserted. As a consequence, carrying out dark field imaging in single focus mode will be strongly limited in accessible field of view depending on the contrast aperture used.

Still, the Lorentz stage position offers a possible alternative to this issue as the main objective lens can be strongly excited to create a first intermediate

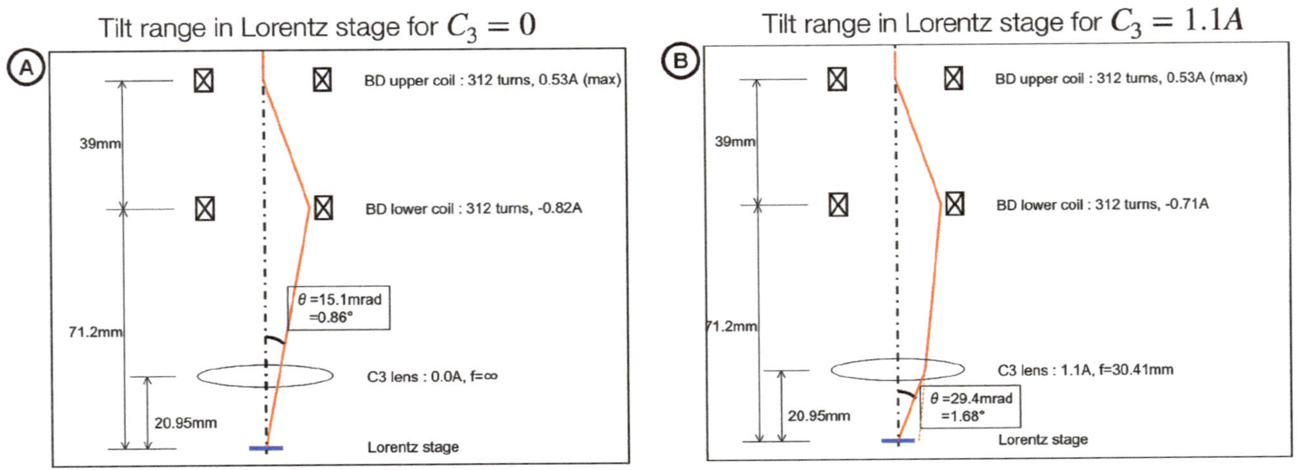

Fig. 1.25 Influence of the C_3 lens excitation on the beam tilt range achievable using Lorentz single or double focus modes.

image in the center of the pole piece gap at $z = 0$ mm, where the sample is normally inserted using the normal stage. This first intermediate image is then focused in a second image located in the standard SA plane position thanks to the objective lens strength and helped by the TL_{11} lens located at $z = 53$ mm as shown using the axial magnetic field curve in Fig. 1.24B. Paraxial rays of this optical mode called "double focus" mode are reported also in Fig. 1.24B. This mode can be set using a strong excitation of the objective lens around 12.8 A, instead of 12.2 A used in normal TEM mode when the sample is inserted using the normal stage. Applying such strong excitation to the main objective lens allowed us to focus the principal ray in the contrast aperture position at $z = 0$ mm. Now that the contrast aperture can be properly inserted in the aperture-stop plane it becomes possible to perform dark field imaging in Lorentz imaging without field of view limitation coming from the aperture, and using beam tilt range going to almost 30 mrad which fulfill all the required specifications in terms of Lorentz imaging for DFEH.

However, Lorentz double focus mode can only be implemented sacrificing the image resolution. Indeed, due to the strong excitation of the main objective lens imposed by the high object plane distance, the aperture aberration increases dramatically to 241 m. This aberration can be decreased using the B COR system, but not sufficiently to reach subnanometric resolution. Together with the chromatic aberration, these aberrations limits the double focus Lorentz imaging mode to 1.8 nm resolution top. The corrected double focus mode still remains in aplanatic condition with a final magnification in the SA plane of 3.8. This value is smaller than the magnification achieved in single focus mode allowing a larger hologram field of view but deteriorating the reconstructed phase resolution. This is the reason why the poor resolution of Lorentz double focus imaging was never really an issue.

To conclude this part dedicated to I2TEM Lorentz imaging, Fig. 1.26 reports all the important numbers associated to the three specific Lorentz imaging modes available using this instrument as well as for the normal imaging mode called "high resolution mode." All these numbers are given for 300 kV of acceleration voltage without B COR system, and with B COR system optimized to perform aplanatic imaging.

1.3.2.4 Electron holography using multibiprism optics

As already described in detail, an off-axis electron hologram is formed by overlapping an image wave Ψ_0^{image} transmitted through a sample with a reference wave Ψ_0^{vacuum} passed through the vacuum area outside the sample (Völkl et al., 1999).

Observation modes	Aberration	Stage	Focusing	Cs	Cc	Resolution	Mag. at SA	Dark field
OBJ off observation Lorentz mode	Uncorrected	Standard	Int 1	770 mm	50 mm	1.0 nm	-	NA
	Corrected (Aplanatic)	Standard	TL11	0 mm	164 mm	0.56 nm	8.8	NA
Double stage (single focus Lorentz mode)	Uncorrected	Upper	Objective (weak)	3.2 m	63 mm	1.43 nm	3.8	NA
	Corrected (Aplanatic)	Upper	Objective (weak)	0 m	115 mm	0.48 nm	13	NA
Double stage (double focus Lorentz mode)	Uncorrected	Upper	Objective (strong)	241 m	1.1 m	4.2 nm	4.1	Available
	Corrected (Aplanatic)	Upper	Objective (strong)	0 m	1.7 m	1.8 nm	3.8	Available
High resolution mode	Uncorrected	Standard	Objective	2.3 mm	2.4 mm	0.24 nm	61	Available
	Corrected (Aplanatic)	Standard	Objective	0 mm	3.7 mm	0.084 nm	99 (Changeable)	Available

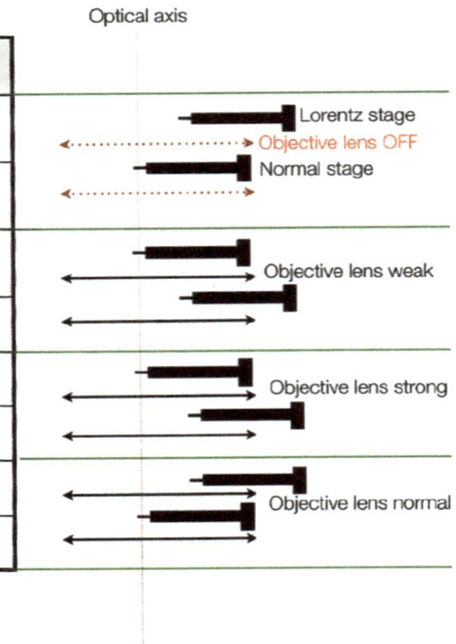

Accelerating voltage : 300kV
Pole piece : Wide gap

Fig. 1.26 Summary of the optical properties offer by the four I2TEM imaging modes either noncorrected or using the aplanatic corrected BCOR properly aligned. From *top* to *down*: Lorentz TL_{11} mode with sample inserted inside normal stage, Lorentz single focus mode, Lorentz double focus mode and finally the normal (or high resolution) mode.

The two important parameters s and W determine respectively the spatial resolution of a reconstructed phase, and the observable size of the specimen inside the hologram. An ordinary optical setup of electron holography used one electron biprism as reported in Fig. 1.16A and is unable to control the two parameters independently. Moreover, the final reconstructed phase is affected by the wire Fresnel fringes superimposed to the hologram.

Carrying out off axis electron holography using two biprisms allowed to get rid of all these artefacts (Harada et al., 2004). The general configuration is reported in Fig. 1.27A. The first biprism (BP_1) is located in the standard position, i.e., after the main objective lens, and before the I_1 lens, and a second one (BP_2) is inserted between I_1 and I_2. BP_1 is excited to generate a shadow area after I_1. BP_2 has to be located within this shadow, and the final hologram created by the superposition of the left and the right wavefronts (respectively blue and red trajectories in Fig. 1.27A) is obtained by polarizing BP_2. By positioning BP_2 in the shadow, it becomes possible to remove all the Fresnel fringes from the BP_2 wire. Additionally, the I_1 lens is set to conjugate the BP_1 plane position with the final hologram plane below BP_2. This allowed to get rid of BP_1 Fresnel fringes as well. In that case, the image plane of the objective lens has to be adjusted to superimpose it with the BP_1 plane, slightly above the standard SA plane.

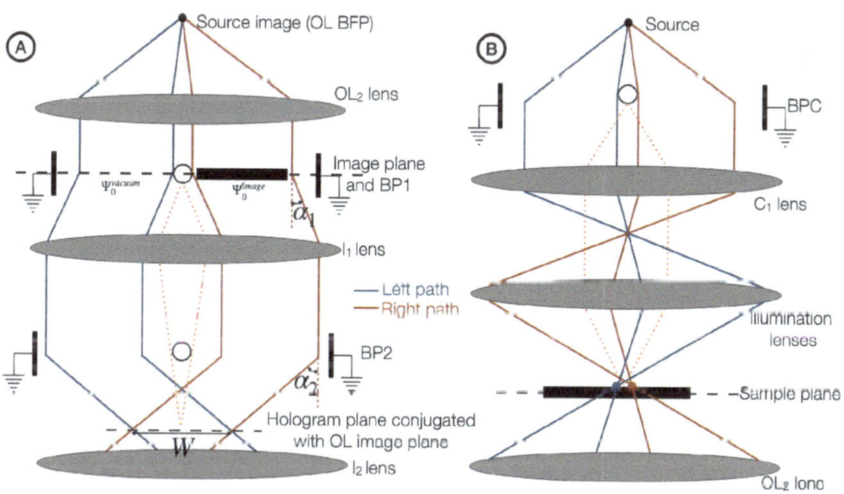

Fig. 1.27 Two multibiprism optical configurations offer by the I2TEM instrument. (A) The double biprism configuration. (B) The split illumination configuration with a *BPC* located below the electron gun and before the illumination lenses.

Moreover this optical configuration, called double biprism electron holography, allowed to set independently *W* and *s*. Indeed, as demonstrated by Harada et al. from HCRL (Harada et al., 2004), *W* is independent from the upper deflection angle α_1 set by the BP_1 voltage. Furthermore, as BP_2 is located very close to the crossover point of the I_1 lens, *s* is also independent from the lower deflection angle α_2 set by the BP_2 voltage. This last condition can be perfect only if BP_2 is exactly located at the crossover position. Still, in the real situation BP_2 is close enough to consider *s* sufficiently independent from α_2. Off axis holograms acquired using the double biprism configuration are reported in Fig. 1.28 showing the flexibility offers by this new optical configuration. We can also appreciate their good quality due to the removal of the Fresnel fringes. Fig. 1.28A displays the effect of BP_2 voltage, increasing the hologram width *W* independently from the interfringes distance *s*. Similarly, Fig. 1.28B reports the effect of BP_1 voltage leading to an increase of *s* keeping the width *W* unchanged.

Thanks to another electron biprism located in the illumination system after the electron source and before the C_1 lens, the I2TEM can be used to implement split illumination configurations (Tanigaki et al., 2012). A schematic of one possible optical configuration is reported in Fig. 1.27B where the illumination lenses have been set to create a pure split shift of

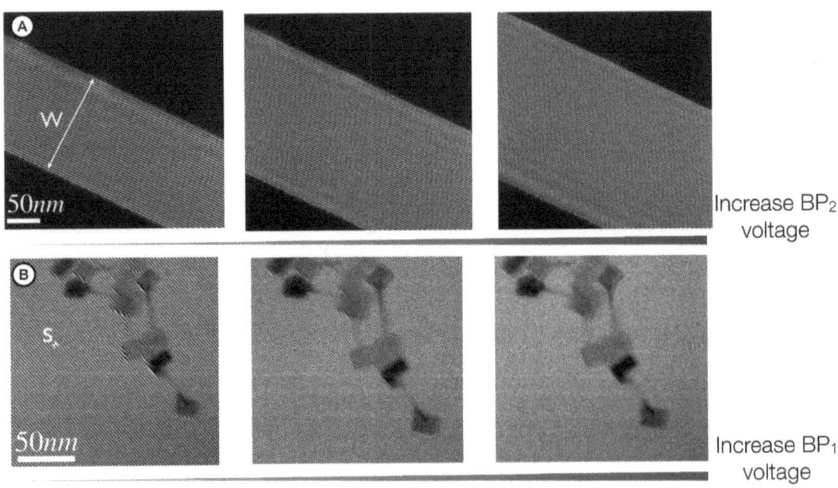

Fig. 1.28 Double biprism electron holograms acquired in the vacuum (A) or with a sample of iron nanocubes (B). (A) Evolution of the hologram field of view *W* with the BP_2 voltage maintaining *s* unchanged. (B) Evolution of the hologram interfringes distance *s* with the BP_1 voltage maintaining *W* unchanged.

the beam in the sample plane (see Section 1.3.2.5). In the literature split illumination is usually used for off-axis electron holography using a reference wave located far away from the sample area (Tanigaki et al., 2012, 2014). This is particularly interesting for magnetic imaging where the magnetic flux lines can disturb the reference wave located close to the sample (Völkl et al., 1999). The condenser biprism (*BPC*) can then be used to select a reference electron wave free from the influence of the sample magnetic flux lines, and the imaging biprism is finally used to perform the final superimposition of this reference wave with the imaging one, either in one (BP_1) or double biprism configuration (BP_1 and BP_2).

In the following I will briefly describe the development of two specifics split illumination configurations applied to strain measurement either using CBED or DFEH methods.

1.3.2.5 Observation using split illumination

Carrying out split illumination using a *BPC* located before the illumination lenses offers a wide range of possible optical configurations. Indeed, by setting properly C_1, C_2 and C_3 lenses we can either create a pure split-shift or a pure split-tilt configuration (Houdellier et al., 2015b; Röder et al., 2016). The first one can be very interesting for many applications such as electron energy loss spectroscopic measurements on nanooptics samples that we have recently started. I will present here it's implementation for strain measurement using CBED. The second one allowed to perform DFEH measurements using a pure unstrained reference beam taken in the vacuum where no diffraction beam can be observed.

1.3.2.6 Development of pure split-shift configuration: Application to splitting CBED

By activating the *BPC*, electron trajectories that passes on each side of the wire will be bend in opposite directions. The further propagation through the illumination system transfers the initial relative tilt into a mixture of related beam shift and tilt. However, due to the particular position of the BPC, the illumination conditions (convergence angle, shift, tilt, probe size) on the sample, which is inserted in the normal stage, will strongly depend on the three condenser lenses strength.

Our goal is to create a splitting CBED pattern (SCBED) by splitting the incident illumination cone in two half-cones of equal size and same semiconvergence angle. To achieve that, the condenser illumination system is set to only shift the two illumination half-cones without introducing additional tilt in the sample plane. This condition is mandatory as the SCBED pattern should contain each half-disks (transmitted or diffracted) of exactly the same diameter (i.e., corresponding to the exact same semiconvergence angle) in the BFP of the objective lens. This configuration is called "pure split-shift" in the following. By using this "pure split-shift" condition on the sample plane, the two half-CBED patterns will be simply added in a single pattern without introducing any shift in the diffraction plane and therefore without any overlapping or separation between the two half-disks of each transmitted and diffracted beams (Houdellier et al., 2015b).

A SCBED pattern is then obtained in the BFP. It consists of disks formed by two half-disks originating from two areas separated by the focused image of the biprism wire. As in standard CBED, each point in the transmitted SCBED disk corresponds to a different incidence of the convergent beam. However, if in standard CBED this intensity distribution is obtained on a single crystalline area, in SCBED it corresponds to the intensity distribution obtained on two areas probed by the two beams respectively denoted Right (R) and Left (L) in the following (see Fig. 1.29). The intensity mapping in each half disk R and L will therefore correspond to the variation of the diffracted intensities as a function of the incident beam direction and obtained in the crystalline regions R and L respectively.

To determine the correct lenses excitations fulfilling the requested "pure split-shift" conditions, we have developed a first paraxial simulation code using MATLAB environment (Hawkes and Kasper, 2018; Houdellier et al., 2015b; Röder et al., 2016). Fast numerical optimization of the trajectories is obtained by using analytical lens field models such as Glaser bell shaped axial magnetic field for each condenser lenses (Glaser, 1952). The marginal electron trajectory starting from the source is finally calculated by Runge–Kutta integration of the standard paraxial equation (Orloff, 2009). As reported in Fig. 1.29A, the "pure split-shift" condition is obtained focusing the *BPC* plane onto the front focal plane (*FFP*) of the objective lens. This leads to a "pure split-tilt" at the *FFP*. Furthermore, to achieve a convergent beam, the image of the source is required to be focused on the object plane. C_1 and C_2 excitations are numerically determined to fulfill these conditions. Fig. 1.29A also shows a detailed ray diagram of the objective lens area (*OB*) for two biprism voltages U_B. The shift between the right and the left parts of the beam (*R* and *L*) can be clearly identified without any tilt contribution.

Various combinations of condensers lenses currents can be used to perform SCBED. The configuration chosen will depend on the shift range

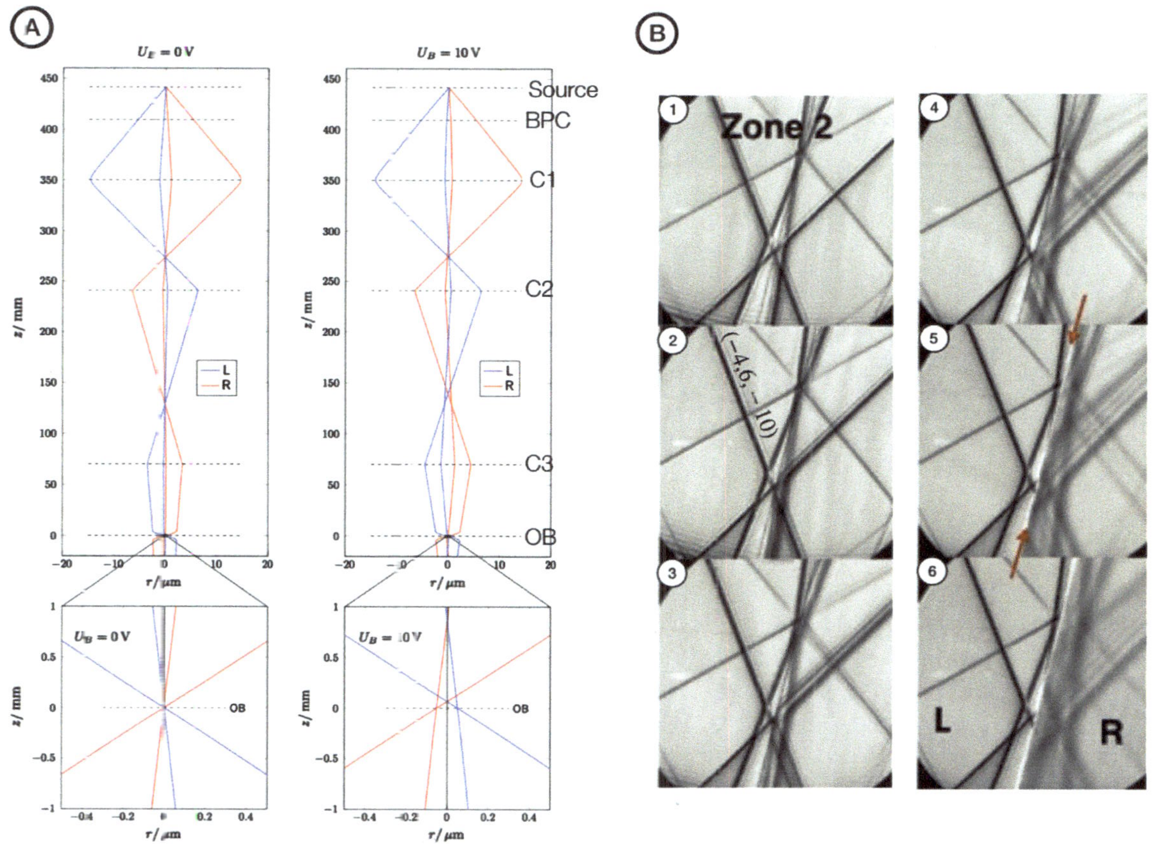

Fig. 1.29 See figure legend on next page.

required for the analysis. The curves used to choose the illumination lenses excitations depending on the shift range can be found in Houdellier et al. (2015b).

SCBED patterns recorded in the Si substrate of a sample constitutes by 20 multilayers of $Si_{1-x}Ge_x/Si$ epitaxially grown on a *Si* substrate, and prepared using FIB system, are reported in Fig. 1.29B. The FIB sample geometry was the same as the one described in Fig. 1.13, with only two constant thickness areas of $t_1 = 145\,\text{nm}$ and $t_2 = 200\,\text{nm}$ called respectively zone 1 and zone 2. SCBED patterns have been acquired along the [750] quasi-kinematical ZAP in the zone 2. The BPC voltage was varied between 0 and 20 V, which corresponds to a maximum shift of 200 nm, as the C_3 strength was set to allow 10 nm/V shift using the *BPC* voltage. Aligning the split-shift direction perpendicular to the interface between the substrate and the multilayers, we can observe the HOLZ line broadening coming from the inhomogeneous strained area on the right side of the SCBED pattern (*R* in Fig. 1.29B) as the *BPC* voltage increases, the associated half spot on the sample becoming closer and closer to the epitaxial interface. All the left half disks of the SCBED patterns (*L* in Fig. 1.29B) remain unchanged and can be used to determine the reference high voltage of the microscope.

Quantitative analysis has been performed for the SCBED pattern recorded with a *BPC* voltage of 10 V. By fitting the HOLZ lines position in the left part of the disk with the one determined using quasi-kinematical simulations, we have measured an acceleration voltage of $300.47 \pm 0.1\,\text{kV}$. Quantitative strain analysis has been performed by fitting the $(-4, 6, -10)$ HOLZ line rocking curve broadening extracted from the right part of the SCBED disk. As previously described, strain state simulations were performed using finite element modeling to take into account thin lamella

Fig. 1.29 (A) Paraxial ray tracing (R and L beams) of the I2TEM condenser system for a pure split-shift set-up with $C_1 = 2.7\,\text{A}$, $C_2 = 2.3\,\text{A}$ and $C_3 = 0.89\,\text{A}$ excitations. Detailed of the paraxial ray paths (R and L) in the objective lens area are reported in the two *bottom boxes*. The BPC voltages U_B are 0 V (*left*) and 10 V (*right*). *z* and *r* define respectively the length of the electron optical column, and the lateral dimension perpendicular to the electron beam. (B) [750] zone axis SCBED transmitted disks observed in the Si substrate of a sample constitutes by 20 multilayers of $Si_{1-x}Ge_x/Si$ deposit on *Si*. The patterns have been recorded on the zone 2 of the FIB prepared sample defined by a thickness $t_2 = 200\,\text{nm}$. The BPC voltages were set between 0 and 20 V (1.0, 2.4, 3.8, 4.12, 5.16, 6.20 V respectively). The *red arrows* indicate the position of the BPC wire in the SCBED disks. *Extracted from Houdellier, F., Röder, F., Snoeck, E., 2015. Development of splitting convergent beam electron diffraction (SCBED). Ultramicroscopy 159, 59–66*

relaxation (Houdellier, 2006). Finally, the variable parameter adjusted to fit the rocking curve, computed using dynamical simulations, was the Ge content x in the multilayer thanks to the Vegard's law. The best fit has been obtained for $x = 20\% \pm 1\%$ of Ge (Houdellier et al., 2015b).

While this new split illumination configuration was first applied to strain measurement using CBED, we are now more interested to implement it together combined with EELS and cathodoluminescence measurements. These specific experiments will be conducted by two recently recruited scientists in CEMES Dr. Sophie Meuret and Dr. Hugo Lorenço Martins. My contribution in this case will be limited to training them on the pure split-shift optical mode. But prior to this, it is necessary to improve the optical simulations used to determine the optimized excitations of the illumination lenses, associated to a specific experimental configuration. Indeed, our first software only computes paraxial equations which, by definition, neglects the influence of lenses aberrations. Carrying out splitting illumination under a pure shift condition will be possible for a sets of excitations which will give identical half spots sizes, as far as paraxial solutions are concerned. However, they will not be equivalent regarding the influence of aberrations. For instance, by implementing shifts larger than a few microns we have noticed the contribution of off-axial coma on each half focused probe, even with the aplanatic B COR. This effect decreases the spatial resolution and introduces a distortion of the SCBED disks.

The development of a new software enabling to compute real trajectories (not only paraxial ones) over the full I2TEM optic from the CFE source to the detector has been performed in collaboration with Hitachi and will be described in Section 1.3.2.8. The goal of these developments was to improve our simulations to finally help I2TEM's users to develop new complex interferometric configurations, such as pure split-shift, but specific to their needs. Prior to these developments, we have implemented a second innovative split illumination configuration allowing pure-tilt on the sample plane and called "TRW" briefly described in the following (Röder et al., 2016).

1.3.2.7 Development of pure split-tilt configuration: Application to TRW DFEH

During the postdoctoral visit of Dr. Falk Röder, we have realized a TRW by means of the combined actions of a *BPC*, the illumination lenses and the deflection system (Röder et al., 2016). This configuration offers new possibilities in DFEH in which a tilted reference wave could replace the dark-field reference wave scattered by an unstrained object area $\Psi_{\bar{g}}^{\text{unstrain}}$

(see Fig. 1.16). This allows to realize a well-defined and object-independent plane reference wave in DFEH. Using the I2TEM optic, a TRW has been obtained by means of the *BPC* which acts as a wave front splitter to introduce a relative tilt between the partial waves passing along opposite sides of the biprism wire. In the following, the deflector system is set to generate an absolute tilt, i.e., tilting both waves in the same way, and the illumination lenses act as an amplifier with respect to the tilt angle.

Indeed, within aplanatic condition, the tilt introduced by the *BPC* between the two waves can be increased by the illumination lenses considering the Abbe's sine condition (Gross, 2005). By demagnifying the biprism plane we therefore obtain a magnification of the tilt angle. Since the biprism stability decreases with increasing voltage, we target a tilt angle given by the lowest possible voltage. Furthermore, carrying out optimum DFEH requires parallel illumination conditions and this condition should remains constant under biprism voltage variation. To find suitable condenser lenses excitations, we used the first homemade paraxial simulation software developed within MATLAB and already introduced in the previous part dealing with SCBED (Hawkes and Kasper, 2018).

Fig. 1.30A shows the paraxial simulation of marginal rays starting from the source to the sample plane with *BPC*, illumination lenses and deflectors (upper and lower) set to generate a TRW condition. Such as for the pure split-shift condition introduced previously, the simulations show that for a certain range of C_3 lens excitations, the C_1 and C_2 lenses can be adjusted to fulfill the pure split-tilt condition in the sample plane, also defined by a pure split-shift in the FFP of the main objective lens as seen in the detailed simulations of Fig. 1.30A. All the lenses excitations allowing to generate various kinds of TRW range conditions can be found in Röder et al. (2016). The action of the deflectors is also very important. Indeed, as observed in Fig. 1.30A, the upper dipole centers one beam (the blue one in the figure) onto the optic axis at the pivot plane of the lower dipole. The lower dipole can then align the beam symmetrically with respect to the optic axis. The C_3 lens remains excited in a such way that the virtual sources are exactly imaged onto the *FFP* of *OB*. Since the blue beam (set as the object beam) is focused on the optic axis, no tilt is introduced on the object plane. Furthermore, the shift of the red beam (set as the reference beam) at the *FFP* will tilt this beam at the object plane, while the object beam remains centered. A change in the biprism voltage then requires an adaptation of the deflection coils such that tilt and displacement conditions remain fulfilled.

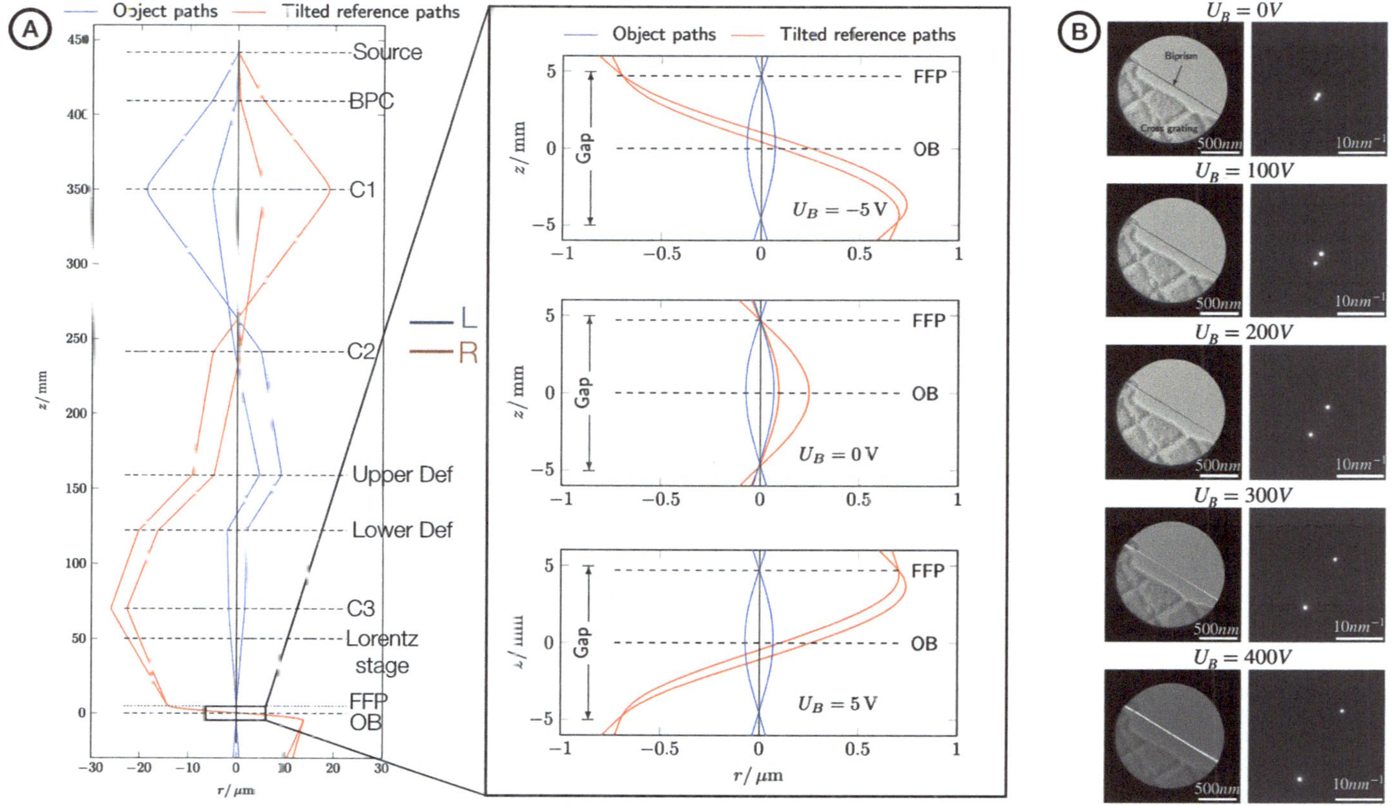

Fig. 1.30 See figure legend on next page.

In order to verify experimentally the TRW conditions we have acquired a series of images and diffraction patterns which are reported in Fig. 1.30B (Röder et al., 2016). As we can observe in the image series (left part of Fig. 1.30B), the biprism nearly stays in-focus for voltages up to $U_B = 200\,\text{V}$ without any mutual tilt between the two parallel half disks. The *BPC* is getting out-of-focus for larger voltages, which may be attributed as an effect of illumination lenses aberrations, as discussed for the SCBED configuration. The diffraction patterns (right part of Fig. 1.30B) indicate that the parallel illumination required is properly fulfilled and show monotonic separation of the corresponding bright field spots with increasing voltage of *BPC*. The shift observed corresponds to the mutual tilt angle introduced between the two half parallel beams. The range of mutual tilt, determined from the shift distances in the diffraction plane, follows a quasi linear dependence.

The method has been finally applied to DFEH. Fig. 1.31 reports the first DFEH experiment realized interfering the dark-field wave with a TRW in the vacuum. The sample used to implement this first proof of concept was an orthorhombic crystal structure of neodymium scandate $NdScO_3$ defined by larger lattice parameters compare to standard silicon ($a = 0.5575\,\text{nm}$, $b = 0.5776\,\text{nm}$ and $c = 0.8003\,\text{nm}$) allowing to generate a closer (002) reflection with $g_{002} = 2.5\,\text{nm}^{-1}$. The sample was prepared by FIB. The specimen was first oriented into a (002) dark field condition. Additionally, as described

Fig. 1.30 (A) Paraxial ray tracing of the I2TEM condenser system including the objective lens allowing to generate a tilted reference wave setup with a *BPC* polarized under $U_B = -100\,\text{V}$ and a condenser aperture radius of $10\,\mu\text{m}$ (located slightly above BPC). The gap's center of the condenser lenses (C_1–C_3), the upper and lower dipoles of the deflection system (upper Def and lower Def), and the main objective lens (*OB*) are marked by *dashed horizontal lines*. The front focal plane (*FFP*) of *OB* is drawn by a *dotted line*. The normal stage is located at *OB* and the Lorentz stage is also reported. The condenser lenses are excited by $C_3 = 0.636\,\text{A}$ and the corresponding currents of the first illumination lenses are $C_1 = 2.6\,\text{A}$ and $C_2 = 2.3\,\text{A}$. Inside the *black square* are reported the paraxial trajectories through the objective lens under variation of the *BPC* voltage ($U_B = -5\,\text{V}$; 0V ; 5V). The pole piece gap of *OB* is marked by a *black arrow*. The beams are focused at the *FFP* where pure split-shift is generated allowing to provide parallel illumination at the center of *OB* with a pure split-tilt condition. (B) Images (*left*) and diffraction patterns (*right*) for a TRW condition obtained under various *BPC* voltages for a C_3 lens excitation of about 0.65 A. Up to 200 V, the center of the biprism wire stays in focus (*black line*). For higher voltages, aberrations of the condenser system introduce a defocus of the biprism. Diffraction patterns show the relative tilt between the two waves. *Extracted from Röder, F., Houdellier, F., Denneulin, T., Snoeck, E., Hÿtch, M., 2016. Realization of a tilted reference wave for electron holography by means of a condenser biprism. Ultramicroscopy 161, 23–40.*

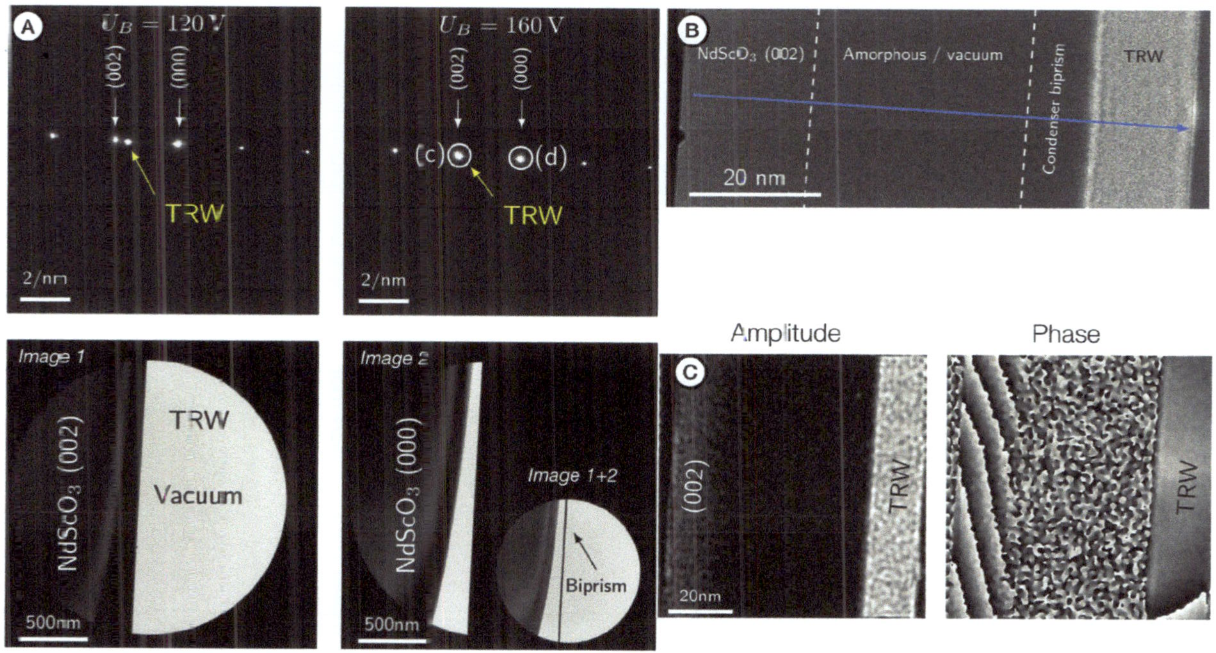

Fig. 1.31 See figure legend on next page.

in Fig. 1.30, both beams were tilted allowing to reach the (002) Bragg angle. Carrying out the TRW set-up, it becomes possible to obtain a tilted reference wave in the vacuum which is at the same time parallel to the (002) scattered wave and to the optic axis. Increasing the *BPC* voltage to $U_B = 160\,V$, the tilted reference wave spot coincides with the (002) reflection as shown in Fig. 1.31A. Inserting a 10 μm contrast aperture around the (002) reflection yields a dark-field image of the object next to the intensity of the tilted vacuum wave as seen in the image below the diffraction pattern in Fig. 1.31A. The superposition of both sides of the images (the dark field on the left with the TRW on the right) is obtained thanks to the BP_1. This superposition have produced the hologram reported in Fig. 1.31B. On the right part, we obtain the superposition of the TRW with itself which will be the reference for the phase analysis, performed using standard Fourier methods previously described. The dark-field hologram created by the superposition of the (002) diffracted wave and the TRW is located in the left part of the hologram. In this region, the amplitude of the hologram fringes as well as the phase relative to the TRW could be reconstructed for the first time, as reported in Fig. 1.31C. It shows the feasibility of TRW under dark field imaging conditions (Röder et al., 2016).

Unfortunately, following this first proof of concept obtained thanks to a well-adapted test sample (due to his large lattice parameters), we have never succeeded in applying this method to a sample of interest such as the stressed microelectronic device used to develop DFEH in the first place.

Fig. 1.31 (A) Diffraction patterns of neodymium scandate oriented in systematic row condition along *c*-axis including the spot of the tilted reference wave for *BPC* voltages of $U_B = 120\,V$ (left) and $U_B = 160\,V$ (right). At 160 V the location of the tilted reference wave coincides with the (002) reflection. The white circles indicate the position of the objective aperture (10 μm) inserted to perform for dark (image 1: left) and bright field imaging (image 2: right). *Inset* in the bright field image on the right, shows the image including both reflections using a larger objective aperture (image 1 + 2). The location of the *BPC* wire can then be observed. (B) Dark-field hologram of NdScO3 using the (002) superimposed with the reference wave tilted in the same direction. The blue arrow indicates the direction perpendicular to the fringes. (C) Amplitude (left) and phase (right) reconstructed from the complete hologram (partly shown in B) with a 2 nm spatial resolution. The bright areas in the amplitude indicate the areas where sufficient fringes contrast can be observed allowing the phase to be retrieved. The phase difference between the (002) dark-field wave and the TRW can be clearly seen on the left part of the phase map. *Extracted from Röder, F., Houdellier, F., Denneulin, T., Snoeck, E., Hÿtch, M., 2016. Realization of a tilted reference wave for electron holography by means of a condenser biprism. Ultramicroscopy 161, 23–40.*

The primary reason stems to the illumination lenses aberrations which are not taken into account in the paraxial simulations. Looking at the paraxial simulations, many kind of configurations are indeed available using I2TEM, but many of these configurations are not feasible practically due to the aberrations contribution which significantly reduces the degree of coherence of the beam and hence, destroys the hologram contrast. Implementing TRW–DFEH on a real silicon based device was possible optically, but the remaining degree of coherence was not sufficient to observe any fringes.

Finally, as for the SCBED configuration, even if the TRW configuration is available using the I2TEM, its optimization requires a more complete simulation enabling the calculation of real trajectories to take into account the nonlinear effects of the lenses. More generally, a complete simulation should benefit all types of interferometric configurations implemented using the I2TEM. The wide optical flexibility offered by the I2TEM instrument is indeed its main weakness. Implementing a specific optical configuration in I2TEM, such as SCBED or TRW–DFEH, requires a deep knowledge of all the possible modes of operation to select the one which is feasible practically. This development direction has been taken in deep collaboration with the HHT factory in Naka and was the main work of Yudai Kubo, a HHT engineer who visited CEMES during 1 year (Kubo et al., 2017). The work of Kubo-San has been then improved during the PhD thesis of Dupuy (2021). I will now describe briefly these developments and their current status.

1.3.2.8 Optimizing I2TEM configuration thanks to full trajectory simulations

Optimizing advanced configurations using I2TEM, ideally requires simulation of the complete electron trajectories within the microscope for any kind of situations. This will give, prior to any experiment, the optimal conditions (i.e., excitations of lenses and deflectors as well as extraction, gun lens and acceleration voltages for the gun part) which should be used experimentally to obtain the desired optical parameters (shift distances, tilt angles, magnification, etc.). The first model used to implement the SCBED and TRW configurations previously described, is limited to paraxial solutions and to simple Glaser axial magnetic field lens models (Glaser, 1952). The paraxial approximation is nevertheless useful to obtain a first estimate of the desired configuration. Among all the possible paraxial solutions, the realistic ones will finally have to be discriminated using a more complete software capable of determining the real trajectories through the whole instrument.

Hence, this will require two specific developments:

1. Developing a real tracing simulation with which the complete trajectories from the source to the detector plane can be extracted. The full magnetic field distribution within the lenses, or electric field within the gun, should be simulated starting from the real design of each component.
2. Improving the existing paraxial simulation and extend it to simulate paraxial solutions over the complete instrument, from the source to the detector. The axial magnetic fields should be extracted from the real components and not approximate using simple Glaser bell shaped model as before.
3. Both simulations should be included within a common software designed to optimize the visualization of these complex trajectories over the complete instrument taking care of the large size differences (length of the column vs tip area for instance)

As briefly introduced at the beginning of this document, describing electron trajectories inside static electromagnetic fields can be performed using the language of optical science thanks to the Hamilton's characteristic function derived from the Lagrange variational principle (Gross, 2005; Rose, 2009). Indeed, a characteristic function, usually the angular one is the most appropriate to describe an optical system, can be developed in power series thanks to the series expansion of the electromagnetic field. For instance in a round optical system such as a magnetic lens, the development of the angular characteristic function, also called angular eikonal function, will contain only even order terms (Rose, 2009). The derivative relatively to the object side or image side angles (for the angular eikonal) of the second-order term will describe paraxial properties of the system while the derivative of the fourth-order term will give access to primary Seidel aberrations, sixth to secondary aberrations, and so on (Hawkes and Kasper, 2018). This first way of describing the real electron trajectories inside a static field by separating them in terms of paraxial (linear) properties perturbed by aberrations (nonlinear) will be called the **optical method**. This method was developed soon after the birth of charged particles optics as it gives relatively simple analytical equations for each contribution (Glaser, 1952; Grivet, 1972). For instance, in the case of single round magnetic lens, the paraxial equation is a time-independent second-order derivative equation depending simply on the axial magnetic field distribution and is given by:

$$r'' + k^2(z)r = 0 \tag{1.60}$$

where the derivative is performed relative to z: $r'' = d^2r/dz^2$. The term $k^2(z) = \frac{(Q^{\text{ref}})^2}{8E_0^{\text{ref}} m^{\text{ref}}} B(z)^2$ where Q^{ref}, E_0^{ref}, m^{ref} are respectively the charge, the kinetic energy and the mass of the reference particle which defines the optic axis and $B(z)$ is the distribution along the propagation direction z of the lens axial magnetic field. The propagation variable is defined by $w = x + iy = re^{i\theta(z)}$ where $\theta(z)$ is the Larmor rotation (Grivet, 1972; Hawkes and Kasper, 2018). More details on the origin of this equation will be given in Section 3.2.3 dealing with the description of the research project.

The two independent solutions of this equation (usually principal and marginal rays) don't have simple analytical solutions, but they can be easily computed using standard numerical methods such as Runge–Kutta ones (Kutta, 1901). These equations were the one used to compute the paraxial solutions of the SCBED and TRW configurations within MATLAB. Analytical expressions of aberrations are also given by the optical method in the form of integral containing the paraxial solutions, given by the paraxial equation, and the power series components of the electromagnetic field (Hawkes and Kasper, 2018; Szilagyi, 1988). These integral equations don't have simple analytical solutions as well, but can also be computed very easily using routine numerical methods such as Simpson rules (Orloff, 2009). Computation of paraxial solutions as well as aberration calculation using aberration integrals are performed by any software based on the optical method such as Munro's electron beam software (MEBS) or electron optics design (EOD) (Chu and Munro, 1982; Lencová and Zlámal, 2008). These softwares are used to designed new single optical components, or new instruments. In that case the optical method is the most relevant. Indeed, computation of all optical properties is then very fast which is a major positive point with regards to the optimization of optical system geometry. The computation speed also explains the attraction of this method in the past when computer systems were slower than today. Indeed, the following method that I will describe requires higher computational time but offers a more direct interpretation of the real effects of the lenses on the trajectories, especially for TEM microscopists who are not used to manipulating charged particle optics concepts.

Hence, this second method to study the electrons trajectories is indeed more "brutal" and relies on the modern capabilities of computers. The movement of a charged particle inside an electromagnetic field can be simply described using Newton's equations and the Lorentz force (Yavor, 2009; Wollnik, 2014). Of course these time-dependent equations don't have analytical solutions, but they can be also computed directly using numerical methods such as Runge–Kutta ones, without separating them in power series like in the optical method (Kutta, 1901). Calculating one electron trajectory inside an electromagnetic field using step by step numerical integration of Newton's differential equations will reveal the real trajectory inside a lens, or a full instrument, depending on the initial conditions of this particle. However, this will not directly give usual optical informations on the system (paraxial or aberrations). Many trajectories will be needed, requiring a lot of computing time, to obtain such information. Moreover, the true paraxial properties will never be easily accessible using these calculations. Indeed, only real trajectories are computed and they are always, by definition, formed by an intrinsic combination of linear and nonlinear contributions. Separating them is more complicated and requires a much longer computing time than the optical method. We will call it **the mechanical method** and it is sometimes called **real tracing method** (Lencová and Zlámal, 2008). Carrying out real tracing is also possible with EOD for instance, but the method is very efficient using SIMION software with which these trajectories can be computed over a whole instrument with many components (Dahl, 2000). SIMION also offers coding capabilities thanks to the internal LUA language and many dedicated variables which can be easily manipulated to perform in-flight calculations or optimization (Ierusalimschy, 2003). The strength of SIMION lies in these coding capabilities allowing us to extract additional information from the real calculated trajectories depending on the user's needs.

For both methods, prior to the determination of paraxial solutions and aberrations or the real trajectories, we have to calculate the electric or magnetic fields inside a lens. These fields will be given by the geometrical shape of the electrodes (electrostatic lenses) or pole pieces (magnetic lenses) and the potentials applied to the electrodes or the excitation of the coil. In the case of magnetic lenses it is also necessary to take into account the permeability of the pole pieces. Let's first consider the case of an electrostatic lens such as the electron gun in a TEM. The electrostatic potential $U(r, z)$ can be determined everywhere by solving Laplace's equation subject to the boundary conditions. These conditions are given by the applied voltages on the various electrodes and at infinity (Orloff, 2009). Because the lens electrodes may have arbitrary shapes it is almost always impossible to do this analytically, and there are number of numerical ways of solving the problem. Three families of methods have been found useful for this purpose: the finite-difference method (FDM), the finite-element method (FEM), and the boundary-element (or charge density) method (BEM) that I will not describe here as we have never used it during our previous work (Hawkes and Kasper, 2018). I will however briefly describe it in Section 3.2 dedicated to computational methods within the research project. In the FDM the space in which the potential field is to be calculated is covered with a rectangular mesh and Laplace's equation is used to relate the values at the nodes. A systematic procedure (relaxation) then enables us to find a set of values that are compatible with the boundary values and *i*th iteration of Laplace's equation (Dahl, 2000). In the FEM the mesh is no longer necessarily rectangular and meshes of very different sizes can be used in different areas, which means that the mesh size can be matched to the rate at which the potential varies—slowly far from the electrodes, for example, and rapidly close to areas where the shape changes, such as corners (Chu and Munro, 1982; Lencová and Zlámal, 2008; Orloff, 2009). Once again, the values at the nodes are related, this time with the aid of a variational approach (see Section 3.2.2), and a solution is found. Both methods are used in different commercial software packages.

SIMION for instance uses FDM methods to determine the potential distribution inside an optical component, while EOD or MEBS used FEM methods (Chu and Munro, 1982; Dahl, 2000; Lencová and Zlámal, 2008). These methods can also be used to deal with magnetic problems. In that case, since the interesting optical region, i.e., the pole piece region, contains no coil windings, and if we assume that the magnetic circuit has a very low reluctance, the magnetic field can be simply expressed as the gradient of a magnetic scalar potential $w(r, z)$ (Yavor, 2009). The same methods can then be applied to map the scalar magnetic potential distribution either by minimizing the variational function F, given by the corresponding Lagrange density Λ (see Section 3.2.2), or by solving the equivalent Laplace's equation subject to the prescribed boundary conditions on the two pole pieces surfaces (Orloff, 2009). This problem becomes then equivalent to a standard electrostatic one. If the coils have to be taken into account, as well as the magnetic saturation effects of the magnetic circuit (pole piece and yoke), the magnetic vector potential $\vec{A}(r, z)$ must be used. The distribution of $\vec{A}$ can be also computed for instance by minimizing the new variational function F extracted from the Lagrange density Λ (see Section 3.2.2) usually using the FEM (Orloff, 2009).

SIMION cannot take into account coils and magnetic saturation effects. The software will always consider a magnetic problem using scalar magnetic potential (Dahl, 2000). On the other hand EOD and MEBS are able to handle a more advanced description of magnetic circuit to take these effects into account. It is indeed very important to take them into account when designing new magnetic components.

The development of a real tracing simulation used to compute any electron trajectories over the whole I2TEM instrument has been performed within SIMION (Kubo et al., 2017). Fig. 1.32A reproduces a screenshot of the SIMION main window where all the I2TEM components have been extracted from "computer aided design" (CAD) models given by the HHT factory. From these CAD models, the active optical components, i.e., pole pieces for the magnetic lenses, tip, extracting and accelerating anodes for the source, are selected using the CATIA CAD-software and are finally exported in a .stl format, a well-known CAD format usually used in the 3D printing community. SIMION can import this .stl format and convert it to the so-called potential arrays (.pa). Potential arrays is a SIMION format where the boundary conditions can be applied on the surface of the electrodes. Solving Laplace's equation by FDM is then possible using the mesh automatically generated by SIMION. The mesh size can be adapted at will using the LUA coding capabilities.

This operation has been carried out for all the I2TEM components which are then assembled in one so-called ion optic bench (.iob) corresponding to the complete instrument, as seen in Fig. 1.32A. The assembling has been performed taking care of the real distances between the different elements given by the HHT factory. Carrying out FDM computation of the electrostatic potential field inside the source was possible only by adjusting the mesh size going from very small close to the tip to a coarser one in the accelerator. Finally, the boundary conditions applied to each lens pole piece have been calibrated by comparing the axial magnetic field obtained with SIMION using scalar magnetic potentials, with the one computed within EOD using FEM taking care of the coil excitation and all the properties of the magnetic circuit. Thanks to this comparison method, it was even possible to take into account in SIMION asymmetric axial magnetic field observed in C_1 and C_2 lenses at high excitation (above 2.5 A), resulting from the magnetic saturation of their permalloy pole pieces. This saturation effect was observed only for C_1 and C_2, the other lenses pole pieces being in pure iron or in permendur for the main objective lens. Once again, all the material properties were given by the HHT factory. A more detailed description of this procedure can be found in Kubo et al. (2017).

Finally, electrons are generated one by one starting from the tip surface with different angles relative to the optic axis. Initially all the electrons are contained in one plane chosen to be xz, where z is the propagation direction. The electrons remain in this plane as far as electrostatic components are concerned, but due to the Larmor rotation introduced by the magnetic

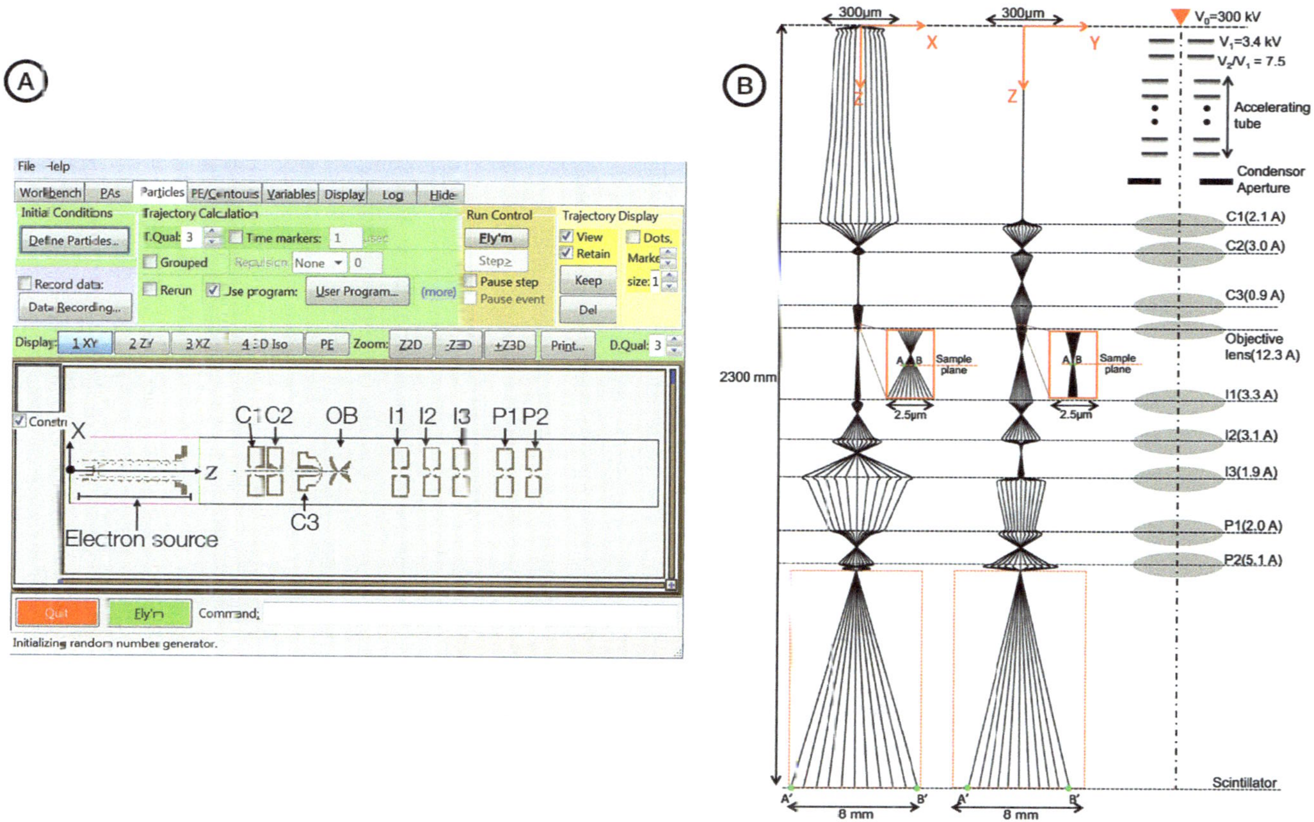

Fig. 1.32 See figure legend on next page.

components, the real trajectories will have projected components in both xz and yz planes (Grivet, 1972). Fig. 1.32B shows the calculation result of nine trajectories from the source to the detector projected in the xz plane (left) as well as the yz one (center). The right schematic simply describes each optical component with their respective excitation used to perform this simulation. The paraxial simulations previously presented in Figs. 1.29A and 1.30A are not described in the cartesian system xyz. Only the paraxial amplitude r given by the complex vector $w = x + iy = re^{i\theta(z)}$ which corresponds to a system representation following the Larmor rotation $\theta(z)$, is shown (Grivet, 1972; Hawkes and Kasper, 2018).

Compared to these first paraxial simulations, our SIMION model allowed us to calculate the real trajectories which intrinsically take into account all the aberrations, even if it can't express them numerically. This simulation has then been used to refine the SCBED configuration taking into account the real lens axial magnetic field distribution in the illumination system, especially the saturation effect of the highly excited condenser lenses, as well as the aberration contributions for large shifts. Simulations carried out for nine trajectories from the source to the sample, and projected in the xz and yz planes, are reported in Fig. 1.33A. Of course, the BPC has been also introduced in the simulation and is polarized using $U_B = 100\,\text{V}$ in that case. The comparisons with the experimental data, especially with the pure-shift condition and the shift distances, are in a very good agreement and can be found in Fig. 1.33B–D. Better optical configurations allowing pure "split-shift" have been derived from this SIMION model. These optimized configurations are now used to perform EELS measurements on plasmonic nanostructures where the two focused half-spots can be located at the same time on two different areas of the structure. In the future, this should allow us to extract new informations on the electromagnetic density of states in nanooptical systems (Lourenço-Martins et al., 2021).

Fig. 1.32 (A) Main window of SIMION where all the I2TEM optical components have been inserted and assembled following the real dimensions of the instrument. (B) SIMION full simulation of 12 electrons trajectories in I2TEM. The *xz* and *yz* planes are reported and enlarged in the direction *x* and *y* by a factor of 800. Inside the *red dashed line rectangle* after the P_2 lens we have shown the electron trajectories in *xz* and *yz* enlarged in *x* and *y* direction by a factor of 40. In the *small red filled line rectangle,* we have also reported the detailed trajectories around the sample plane. *Extracted from Kubo, Y., Gatel, C., Snoeck, E., Houdellier, F., 2017. Optimising electron microscopy experiment through electron optics simulation. Ultramicroscopy 175, 67–80.*

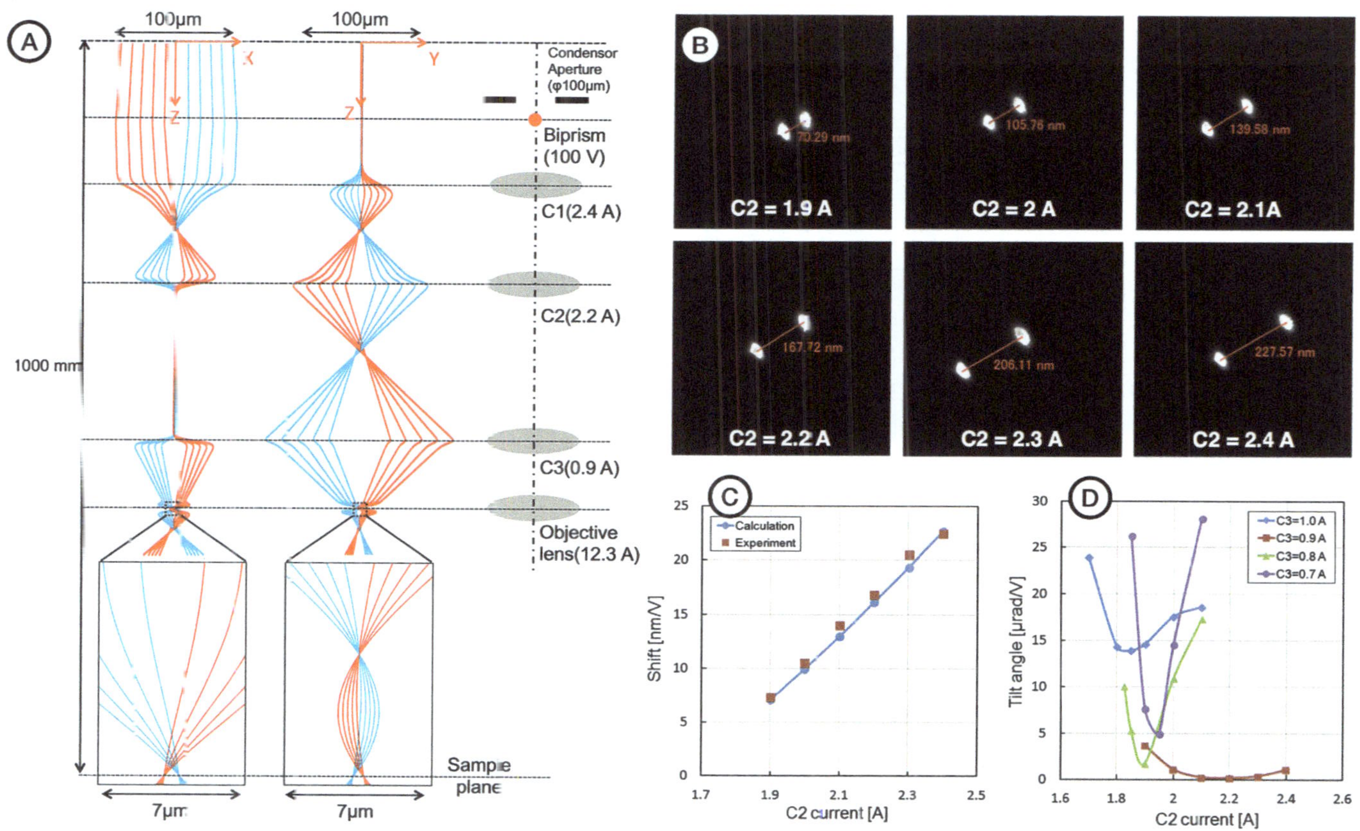

Fig. 1.33 See figure legend on next page.

Additionally to the SIMION real tracing model of the I2TEM, we have also improved the paraxial ray tracing software. In the previous one developed within MATLAB, the numerical solutions of the paraxial equation, computed using Glaser's Bell-shaped axial magnetic field model for each illumination lens, was achieved by an explicit Runge–Kutta algorithm, the formula for which can be found in the Bogacki and Shampine paper (Bogacki and Shampine, 1989). Improving this paraxial simulation first requires to solve the paraxial equation using real axial magnetic field distribution $B(z)$ for all the lenses, such as the one calculated within SIMION. The simulation must be also extended to compute the trajectories inside the source as well as inside all the imaging lenses of the microscope, to generate a complete paraxial simulation of the whole instrument. This extended paraxial model will complement the simulation provided by SIMION. Indeed, paraxial properties cannot be easily extracted using SIMION as already explained. Furthermore, computing a single trajectory from the source to the detector using SIMION requires 2.5 s, while the paraxial simulation of the full column only requires one hundredth of second.

In the PhD thesis of Julien Dupuy, a new semianalytical approach has been developed for the resolution of the paraxial equation and its implementation was achieved using Python scripting (Dupuy, 2021). Briefly, the idea was to mix the discretized axial magnetic field $B(z)$ calculated for each mesh point of the lens model within SIMION together with the paraxial equation given by Eq. (1.60). Indeed, for each mesh area i the axial magnetic field $B(z)$ calculated by SIMION has a constant value B_i. Then for each mesh area i the two independent paraxial solutions, the marginal and principal rays, can be expressed analytically using harmonic oscillator expressions as the paraxial Eq. (1.60) becomes a second-order differential equation with constant parameter. Then, using proper boundary conditions between mesh

Fig. 1.33 (A) SIMION simulation of nine electron trajectories calculated using the SCBED configuration at 300 kV. Only the illumination part of the I2TEM column, in *xz* and *yz* planes, is depicted. *Blue* and *red* trajectories correspond to the electrons crossing on the *right side* and *left side* of the biprism wire, respectively. (B) Experimental SCBED haft focused probes observed at 300 kV and $C_3 = 0.9$ A, using different C_2 lens excitations. (C) Comparison between experimental and simulated shift in nm per biprism voltage between the two half focused probe with $C_3 = 0.9$ A. (D) Effect of C_3 excitation on the split-shift purity. The residual tilt between the two focused half cone is reported as a function of C_2 current and calculated for four values of C_3. *Extracted from Kubo, Y., Gatel, C., Snoeck, E., Houdellier, F., 2017. Optimising electron microscopy experiment through electron optics simulation. Ultramicroscopy 175, 67–80.*

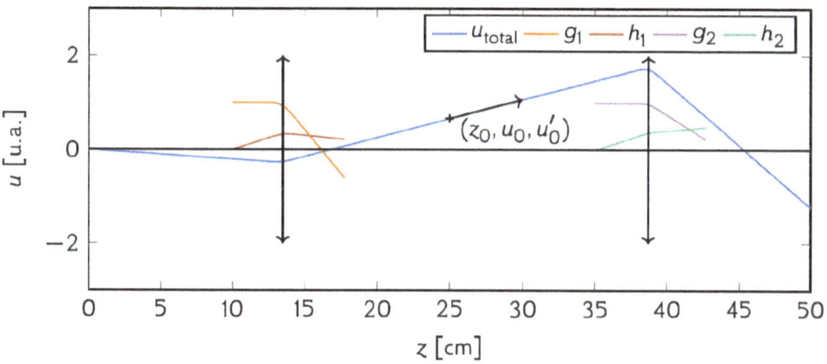

Fig. 1.34 Paraxial solutions calculated using the new semianalytical approach and computed with an homemade Python software, in the case of a simple two lenses system. Once the fundamental rays h_1, g_1 of the first lens and h_2, g_2 of the second lens are calculated, we can obtain the total paraxial solution u_{total} by linear combinations of these pairs of rays in their respective areas. The coefficients associated are determined by maintaining the continuity of u_{total} (in particular at the boundary conditions between the field zones and the free propagation areas), and respecting the initial conditions (z_0, u_0 and u_0'). These boundary conditions result in a system of linear equations that link the coefficients of each of the linear combinations, which can then be solved properly. *Extracted from Dupuy, J., 2021. Contrôle dynamique et optimisation des observations en microscopie électronique en transmission. Université Toulouse 3, Toulouse, PhD thesis.*

areas $i-1$, i, and $i+1$ it becomes possible to compute the marginal and the principal rays for each lenses. An example of simulation is reported in Fig. 1.34 for a two-lens optical system where g and h represent respectively the principal and the marginal ray of the first (g_1 and h_1) and the second lens (g_2 and h_2). Once the two paraxial rays have been computed for each element, the complete trajectory can be easily found using linear combination as reported in Fig. 1.34 where u_{total} represents one paraxial ray of this simple two lenses system. The complete mathematical description of this semi-analytical method used to compute the paraxial solutions thanks to axial magnetic field distribution extracted from SIMION, can be found in the Julien Dupuy's PhD manuscript (Dupuy, 2021).

Finally, as introduced at the beginning of this section, the two simulations have to be included in a single visualization code with a user-friendly vision. The target of such an application will be to help I2TEM users to prepare, prior to the experiment, an optimized optical configuration of the I2TEM depending on the configuration needed. A screenshot of the final visualization code is reported in Fig. 1.35A where the left column corresponds to the paraxial solutions calculated for one possible double biprism

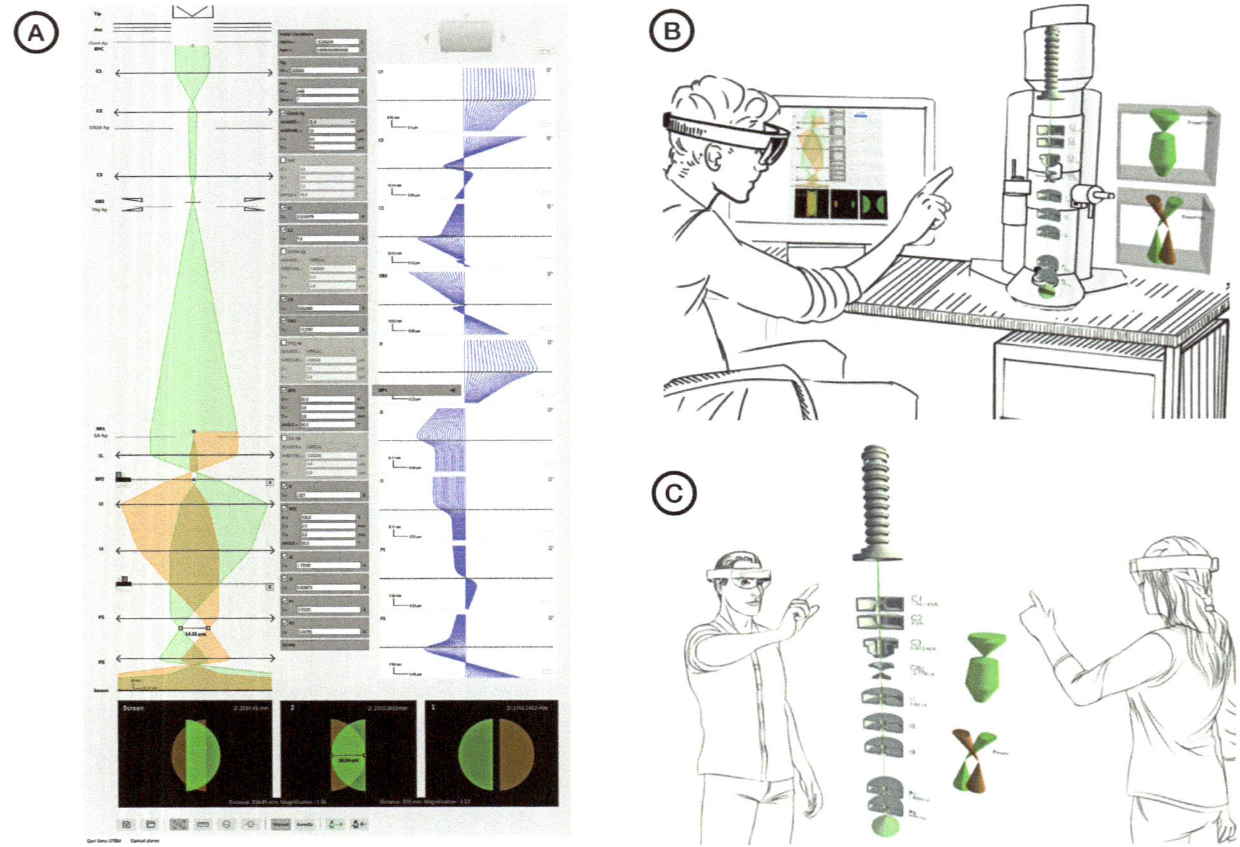

Fig. 1.35 (A) Main window of the I2TEM trajectories visualization software showing (from *left* to *right*) the paraxial trajectories, the excitation values of all I2TEM components used to perform the simulation and the real trajectories given by SIMION. (B) and (C) Two typical situations where the use of mixed reality could be interesting to implement for a better visualization of tridimensional electrons trajectories inside a microscope. *Extracted from Dupuy, J., 2021. Contrôle dynamique et optimisation des observations en microscopie électronique en transmission. Université Toulouse 3, Toulouse, PhD thesis.*

configuration. Various measurement tools (distances, angles, ...) have been also included. The center column is used to define the excitations of each I2TEM elements (currents, voltages, ...) while the right part shows the SIMION real tracing simulation. In that case the trajectories are projected in a plane selected by the user using a tool located at the top. The three squares at the bottom of the software window report the beam shape in three planes chosen by the user and located at three positions perpendicular to the optic axis (z_1, z_2 and z_3). This visualization code has been developed during Julien Dupuy's thesis with the help of two groups of master students and one scientist (Dr. Sylvain Pauchet) from the ENAC (Ecole Nationale de l'Aviation Civile) who worked exclusively on this development during 1 year. More details concerning this software can be also found in Julien Dupuy's PhD manuscript (Dupuy, 2021).

The future direction taken by Dr. Pauchet together with the I3EM group for these specific visualization developments will be to adapt the three-dimensional trajectories visualization using modern augmented reality and virtual reality tools as described in Fig. 1.35B and C.

I think this will be a very interesting direction to explore, but as a nonspecialist I will not participate deeply to these developments oriented to the human–machine interaction. My future prospect will be to improve the SIMION simulation in order to include quantitative aberrations determination. I will also explore the possibility offers by COMSOL multiphysics to improve the computation of the electrostatic field and axial magnetic field distribution of the I2TEM optical elements which may be more relevant than FDM in the case of complex geometry (COMSOL AB, n.d.). At longer term I wish to develop a homemade code carrying out field mapping computation as well as paraxial properties and aberrations determination within one unique language allowing the optimization of an optical system without the interaction of too many software packages such as SIMION, COMSOL, EOD, ... Field computation will be performed using BEM methods, and the code will be written in Python (see Section 3.2). These theoretical developments will be also very important to support the future hardware developments that I wish to carry out, for instance the MOLENS project which deals with the design of a preaccelerated magnetic lens CFEG. They will be described in the last chapter.

Discussion

This chapter addressed one major part of my research activity dealing with the characterization of nanomaterials, mainly strain mapping at the nanoscale, using coherent TEM methods. These methods have been implemented using two kinds of illumination mode on the sample, either coupling the sample plane with the illumination image plane, which allows a very small electron probe to be focused on the sample like in STEM/CBED, or with the illumination pupil plane to generate a parallel beam on the sample like in electron holography. In both cases brightness is the key parameter. Increasing the brightness in an optical system is however impossible. Its value in the different optical planes of an instrument cannot be higher than the one given by the electron source. However, it can be quickly deteriorated by the lenses aberrations either intrinsic, or parasitic introduced by misalignments, instabilities … Following my initial works on the development of CBED and DFEH exploring new kinds of methods to determine the strain state of semiconductor devices, I chose to focus my research on the brightness optimization of coherent TEM methods. Improving the brightness could be implemented following two distinct directions, either by optimizing the optical configuration of a given instrument to minimize the contribution of intrinsic and parasitic aberrations, or directly by increasing the original brightness given by the electron source.

I have first followed the optimization direction which has resulted, thanks to our collaboration with HHT formally started during that time, in the development of the I2TEM instrument. The specifications as well as the optical properties offered by this unique instrument have been described in this chapter. The I2TEM is now a central instrument in CEMES used for various kinds of holography experiments such as DFEH. The development of optimized split illumination configurations, mainly "split-shift" mode, thanks to the full trajectories simulations have opened new perspectives to study optical properties at the nanoscale. These topics are currently under discussion with the LPS in Orsay and the first experiments performed are very promising. In the future I will continue to work following this "brightness optimization" direction with the help of new simulation approaches.

In the next chapter I will discuss the developments performed following the second direction, which is a more instrumental one. Improving the brightness of modern electron sources requires a thorough knowledge of the electron gun technology used in TEM. The types of the electron sources are determined by the physical mechanisms of electron emission from the gun cathode materials. There are essentially three kinds of electron emission in most TEM guns: thermionic emission, field emission and Schottky emission, where Schottky emission can be considered as a mixture of thermionic and field emission. The source based on thermionic emission can generate very intense electron beams but they are brightness limited. On the contrary, sources using field emission are the brightest among all source technologies. Following the second development direction, I have started to work on new field emission technologies trying to improve their intrinsic brightness and also to enable the generation of bright and ultrashort electron pulses, thus paving the way for coherent time-resolved methods using electron beams. I have initiated these developments almost at the same time as we began to install the I2TEM instrument. The whole story really started with the recovering of three old Hitachi HF2000 given by the EPFL in Lausanne and the Physics department of Bristol's university for one symbolic euro.

CHAPTER TWO

Cold field emission electron source: From higher brightness to ultrafast beam

Contents

Si tu es pressé, fais un détour

proverbe japonais

2.1. Introduction

In the first chapter we discussed how coherent methods can be implemented in a TEM if enough coherent current can be transported to the sample plane. For STEM/CBED applications, the produced beam can be focused in a small spot, or in a specified sample illumination form for interferometric applications. We already discussed how we could optimize the beam spatial coherence on the sample plane by adjusting the TEM optical configurations for a given source brightness. We will now detail

ISSN 1076-5670
https://doi.org/10.1016/B978-0-443-19324-8.00002-X

more technically the operation of a coherent source in a TEM and the range of possibilities offer by the modifications of some specific components.

As briefly introduced in the discussion of the previous chapter, several electrons sources are available in TEM: thermionic, cold field emission and Schottky electron sources (Williams and Barry Carter, 2008). In the following parts, we'll go through the basic mechanisms of electron emission from metallic cathodes, which are used in all of these technologies.

The highest occupied electronic energy level inside a material at $T = 0\,K$ is called the Fermi level, E_F (Kittel, 2005). When the temperature T of the system is increased, some electrons can gain energy and occupy higher energy levels. The occupation number $f(E, T)$ describes the probability of having an electron on a state of energy E. It is given by the Fermi–Dirac (FD) distribution (Hawkes, 2009):

$$f(E,T) = \frac{1}{exp\left(\frac{E-E_F}{k_B T}\right) + 1} \quad (2.1)$$

where k_B is the Boltzmann's constant.

As shown in Fig. 2.1 at $T = 0\,K$ the FD distribution is a step function (black curve) that changes with the temperature of the system.

To be emitted in vacuum, an electron located inside a material at the Fermi level requires an extra energy at least equal to the material work function Φ. At high temperature, the F–D distribution includes an appreciable fraction of occupied electronic states with energies above the work function and therefore these electrons are emitted in vacuum. This is the basis of *thermionic emission*, in which electrons are extracted by heating a metallic filament (Williams and Barry Carter, 2008). The case depicted in Fig. 2.1 corresponds to a [310] oriented monocrystalline tungsten (W) cathode defined by $\Phi = 4.25\,\text{eV}$ (Hawkes, 2009). The unit vector $\vec{z}$ is taken perpendicular to the emitting surface of the source.

In absence of applied field the potential is still a step function (black curve) that is modified due to the image charge (red curve). An applied electric field F_{loc} modifies the potential step (blue and magenta curves) following the general expression:

$$V(z) = \Phi - eF_{loc}z - \frac{e^2}{16\pi\epsilon_0 z} \quad (2.2)$$

Thanks to this applied field F_{loc}, the potential in the vacuum follows a linear curve starting from the cathode surface. When the potential barrier between

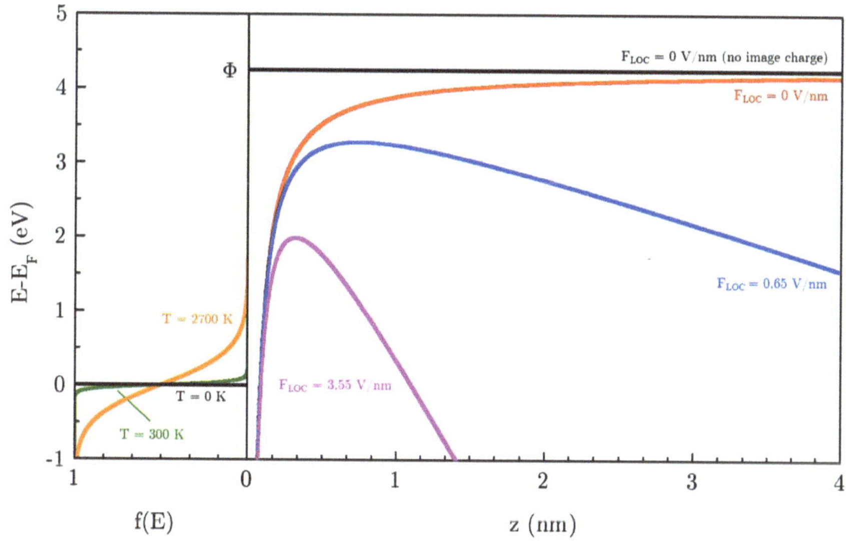

Fig. 2.1 *Left side*: Fermi–Dirac (F–D) distribution at three different temperatures: $T = 0\,K$ (*black curve*), $T = 300\,K$ (*green curve*), $T = 2700\,K$ (*orange curve*) that is the typical temperature used for thermionic emission from tungsten. *Right side*: Potential barrier at a metal–vacuum interface for [310] oriented tungsten ($\Phi = 4.25\,\text{eV}$) for different values of the applied electric field. The *black curve* represents the potential barrier without any applied electric field and neglecting the image charge contribution. The *colored curves* take into account the image charge in three cases: no applied field (*red*), Schottky electron gun $F_{loc} = 0.65\,\text{V/nm}$ (*blue*) and Cold Field Emission Guns $F_{loc} = 3.55\,\text{V/nm}$ (*magenta*). *z* is perpendicular to the metal surface.

the material and the vacuum, around the Fermi level, is thin enough, the electrons can tunnel through it. This is called *field emission*. Additionally, the applied field F_{loc} decreases the exit work function barrier by a quantity:

$$\Delta\Phi = \sqrt{\frac{e^3 F_{loc}}{4\pi\epsilon_0}} \tag{2.3}$$

This is called the Schottky effect. Schottky field emission sources are taking advantage of the combination between field emission and Schottky effect (Swanson and Schwind, 2008).

The selection of a suitable emitter with some specific properties depends on application requirements. For applications where low intrinsic energy width and high brightness is required, the thermal field Schottky emitter (SFE) or the cold field emitter (CFE) are the preferred choice. Each of these two technologies have their own "pro" and "con" which can be found in the literature (Dellby et al., 2011; Fransen et al., 1999b; Hawkes, 2009; Swanson and Schwind, 2008). For instance, it is well known that cold field emission technology is more difficult to implement due to the combination of high voltage and ultra high vacuum mandatory to obtain the necessary current stability (Todokoro et al., 1982). But, in the case of standard TEM/STEM probe current, CFE are intrinsically brighter than SFE (Dellby et al., 2011; Swanson and Schwind, 2009). That's why all the developments performed during my research work, have been always performed on CFE gun technology which I will go over in depth in Section 2.2.

2.2. Principle of cold field emission electron source technology

As described in Fig. 2.2A, FE electrons can be obtained by applying several thousand volts between the tip of a metal needle (FE tip) with a radius

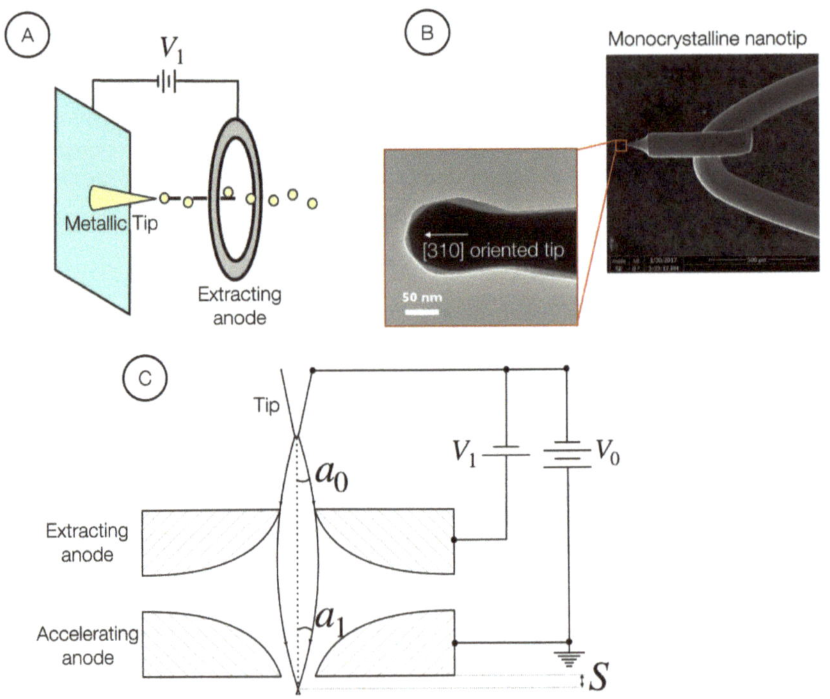

Fig. 2.2 (A) Simplified outline of a cold field emission source. (B) SEM image of a tungsten emitter used in a conventional CFEG. The TEM image of the tip apex is reported in the *red square*. The *W* cathode is oriented along the [310] zone axis. (C) Optimum electrodes shape used in a triode gun calculated by Butler.

of less than 100 nm, and a circular extracting anode. The extraction voltage will now be always expressed using V_1.

The energy spread of the emitted electrons is narrow ($\Delta E = 0.2$–0.3 eV) because the emitter runs at room temperature. The current density J(A/cm^2) is obtained using Eq. (2.4) from Fowler and Nordheim (1928). This density strongly depends on the electric field applied to the FE tip F_{loc} and on the work function of the metal Φ.

$$J = 1.54 \times 10^{-2} \frac{F_{loc}^2}{\Phi . t^2(y)} \exp\left(\frac{-6.83 \times 10^5 \Phi^{3/2} v(y)}{F_{loc}}\right) \tag{2.4}$$

where $t^2(y)$ and $v(y)$ are elliptic functions nearly equal to 1 (Fowler and Nordheim, 1928; Hawkes, 2009). In comparison, the thermionic electron current density is obtained using the Richardson–Dushman equation (Dushman, 1930). The CFE current density (10^4–10^6 A/cm^2) is three orders of magnitude larger than the one obtained using thermionic electrons (1–10 A/cm^2) (Crewe et al., 1968; Hawkes, 2009; Williams and Barry Carter, 2008). The FE source is ideally a point source, and the diameter d_v of the virtual source ranges from 5 to 10 nm because of the small FE tip, which is 1/1000 the source size of the thermionic emission (1–10 µm). Using Eqs. (1.37) and (1.38), we can now fully understand why CFEG is the optimum source for coherent TEM techniques thanks to its small virtual source size and, as a consequence, the maximum coherent current available. If we consider a tip with a hemispherical end, and also assume that electrons will be emitted within a finite voltage range 0 to V_T, then it was demonstrated that the apparent source radius r is approximatively (Crewe et al., 1968):

$$r = R\left(\frac{V_T}{V_1}\right)^{1/2} \tag{2.5}$$

where R is the radius of the tip. This expression is good for a spherical source only, which is a reasonable approximation for FE tip as observed in Fig. 2.2B where $R = 50$ nm. Knowing that $V_T = 0.5$ V and $V_1 = 1$ kV we can find $r = 1$ nm. Under this approximation we can assume that the source size d_v used to calculate the brightness in Eq. (1.38) is then defined by $d_v = 2 \times r$. In practice the source size are bigger than the one obtained using this approximation, but not by an order of magnitude as already mentioned previously.

The following table shows a comparison of emitters characteristics between FE and thermionic electron guns (Hawkes, 2009; Reimer and Kohl, 2008).

	Field emission W [310]	Thermionic emission (W filament)
Source brightness	10^9 A cm^{-2} Sr^{-1}	5.10^5 A cm^{-2} Sr^{-1}
Source size	5–10 nm	1–10 μm
Energy spread	≈0.3 eV	≈2 eV
Temperature	300 K	2800 K
Vacuum pressure	$<10^{-8}$ Pa	$<10^{-3}$ Pa

Typical cathode material in modern CFEG are made with mono-crystalline tungsten (W). The tip is oriented either along [310] or [111] zone axis corresponding to crystallographic orientations with minimum exist work function Φ (see Fig. 2.2B) (Hawkes, 2009). The tip is etched by immersing it in a sodium hydroxide (NaOH) solution and by applying a DC voltage (around 12 V) between the tip and a remote electrode also immersed in the solution (Ibe et al., 1990). Even when employing ultra high vacuum (UHV) environment, one of the challenges in designing CFEG is to avoid emission current instability owing to adsorption of residual gases on the cathode surface (Todokoro et al., 1982). This effect first increases the work function Φ which then decreases the emission current. Then, the diffusion of adsorbed gases and surface sputtering is at the origin of irregular current fluctuations. They can eventually breaks down the cathode by generating a vacuum arc. As a result, the lower the vacuum, the better the CFEG current stability will be. In practice, using standard vacuum technology allowing to reach 10^{-9} Pa range, the emitter has to be cleaned by thermal heating (called flashing) at intervals as short as a few hours (Hawkes, 2009).

The electron gun must have a minimum of three electrodes when using a field emission source in an electron microscope (see Fig. 2.2C) (Crewe et al., 1968). Two voltages can then be applied in order to allow independent control of the emission current and the final electron energy. The voltage V_1 applied between the tip and the extracting anode defines the total emission current and the acceleration voltage V_0 defines the electron energy. Apertures in the first and second anode allow electrons to pass through to form the beam. The distribution of the electric field going through these apertures will be responsible of the focusing effect of this three electrodes system (Hawkes and Kasper, 2018).

Thus, like any electron optical system, this electrostatic lens will have paraxial and aberrations properties. Due to the cylindrical symmetry of this optical system, third order aperture aberrations $(x|aaa) = (y|bbb) = (r|aaa) = C_s$ will be the major limiting factor (Wollnik, 2014). Indeed, this spherical aberration will broaden the source size d_v of the final setup, observed in an optical plane located after the gun. This will strongly decrease the total brightness B_r of the source, and then decrease the coherent current available. That's why, the electron gun should be designed to have minimum spherical aberration, as well as mechanical vibrations, which are the two main reasons behind catastrophic broadening of the final source size d_v (Crewe et al., 1968). Usually a CFEG is designed using the so-called "Butler" electrostatic lens which has an optimum spherical aberration (Butler, 1966). As described in Fig. 2.2C, an FE electron gun with a Butler lens has a FE tip and a diode area for the extraction and acceleration electrodes, with particular surface curvature predicted to reduce the gradient of the electric field along the electron path when the beam passes through the two apertures (Butler, 1966).

In the case of source optic, the aperture aberration coefficient C_s is referred to the emitter plane (Crewe et al., 1968). Hence, the effect of the aperture aberration C_s is to convert an ideal point source into one with an apparent radius of:

$$\Delta r = C_s a_0^3 = C_s (x|x)^3 \left(\frac{V_0}{V_1}\right)^{3/2} a_1^3 \tag{2.6}$$

where a_0 and a_1 are the electrons beam half-angles at the source and image plane respectively (see Fig. 2.2C). The conventional linear magnification in the Gaussian image plane is defined using the Helmholtz–Lagrange paraxial invariant $(x|x)(a|a) = (V_1/V_0)^{1/2}$. Due to the cylindrical symmetry, we know that the linear magnification is identical along any directions perpendicular to the optic axis $(x|x) = (y|y) = (r|r) = m$. The theoretical radius of the spot measured at the Gaussian image plane due only to aperture aberration is then:

$$r_s = m\Delta r = C_s m^4 \left(\frac{V_0}{V_1}\right)^{3/2} a_1^3 \tag{2.7}$$

Finally, the real image size is obtained including the effect of source size using Eq. (2.5) and the diffraction effect considered only on the first diffraction zone:

$$d_{total}^2 = (2 \times r_s)^2 + \left(2 \times mR\left(\frac{V_T}{V_1}\right)^{1/2}\right)^2 + \left(2 \times 0.3\frac{\lambda_1}{a_1}\right)^2 \tag{2.8}$$

where λ_1 is the electrons wavelength in the Gaussian plane. Curves showing theoretical and experimental optical properties of CFEG using Butler lens can be found in the literature (Crewe et al., 1968). In order to have some order of magnitude in mind, let's assume a Butler system with a voltage ratio of 60 given by $V_0 = 60\,\mathrm{kV}$ and $V_1 = 1\,\mathrm{kV}$, and a tip defined by a radius $R = 50\,\mathrm{nm}$ located 10 mm above the extracting anode. This configuration will have a magnification $m = 0.23$ and an aperture aberration coefficient of $C_s = 11\,\mathrm{cm}$.

Sources with a large value of d_v, such as thermionic source, or even Schottky FEG source with $d_v \approx 30$–$60\,\mathrm{nm}$, require the column to be operated in a strongly demagnified mode. This, combined with a high extraction value ($V_1 \approx 4$–$5\,\mathrm{kV}$) results in such a small gun lens aberrations contribution that it can typically be ignored or at most considered a minor contribution to the total column aberrations (Barth and Kruit, 1996). In comparison, for an equivalent probe current, due to the small source size, good CFE performances are achieved using larger value of m than for SFEG, which in turn leads to a larger contribution from the gun lens to the overall aberrations in accordance with Eqs. (2.6) and (2.7). This is a consequence for any CFEG that achieves a large brightness due to its intrinsic small source size as opposed to a large value of angular current density, which is the case for SFEG (Hawkes, 2009). Additionally, the larger value of m used for CFEG explains why more mechanical stability and magnetic shielding are required for the total instrument.

I would like to mention also the contribution of chromatic aberration which can become not negligible when the beam energy V_0 is small, such as for low voltage SEM where V_0 could be in the kV range. In that case, the nonlinear contribution $\Delta x = (x|a\delta)a_0\delta_0$, previously introduced in Section 1.1.3.3, has to be considered in the calculation of the total spot size contribution given by Eq. (2.8). Considering the cylindrical symmetry and the very small object size given by the FE tip, we will only take into account axial chromatic aberration coefficient C_c and neglect chromatic distortion D_c (Rose, 2009). In the same way as for the aperture aberration, the axial chromatic aberration coefficient C_c is referred to the source. That's why, the contribution of axial chromatic aberration to the total beam size in the image plane is calculated using Eq. (1.39) and the Lagrange–Helmoltz paraxial invariant (Crewe et al., 1968):

$$d_c = 2 \times m\Delta r_c = 2 \times m^2 C_c \frac{\Delta E}{V_0} \left(\frac{V_0}{V_1}\right)^{1/2} a_1 \tag{2.9}$$

To sum up this crucial point, we can see that CFEG's greatest strength is also its greatest vulnerability. Indeed, CFEG's small virtual source size leads to a higher column magnification necessary for optimum performance; this, in turn, leads to an increased contribution of the gun lens aberrations which, if not obviated, minimizes the advantage of the high source brightness. One of the key challenges addressed by my research project described in the last chapter will be improving gun lens aberrations to enhance final source brightness.

2.3. 200 keV CFEG of Hitachi: The HF2000

At the heart of the developments on coherent electron sources that I will described in the next parts, is the HF2000. The HF2000 is an Hitachi FE-TEM developed at the end of the 80's that we used to modify and test our instrumental ideas (Isakozawa et al., 1989). Indeed, since 2008 we have collected 4 HF2000 instruments all around Europe. Nowadays, 2 instruments are running in CEMES and can be used to test new hardware, and 2 are used for spare parts.

2.3.1 Multistage acceleration CFEG

Akira Tonomura and his team working in Hitachi Central Research Laboratory (HCRL) developed in 1972 a Butler lens type FE source working at $V_0 = 50\,\text{kV}$ (Tonomura, 1973). They installed it for the first time in a TEM to perform new electron holography's experiments thanks to the high spatial coherence of the CFEG. In a direct continuation, they developed in 1978 a 80 kV FE-TEM with a higher brightness and obtained 3000 lines of Fresnel fringes observed on the defocused image of the condenser aperture hole (Tonomura et al., 1979b). This number must be compared with the 300 fringes obtained with the previous 50 kV source. This allowed them to obtain practical electron holography datas (Tonomura et al., 1979a).

Many remarkable CFE-TEM experiments have been performed with the 80-kV CFE-TEM at HCRL; however, it was difficult to exceed an acceleration voltage of 80 kV with the Butler lens CFEG due to discharge occurring between the emitter and electrodes. The same team in HCRL developed in 1984 a multistage electrodes gun that produced an acceleration voltage of 125 kV and they could also show a significative improvement of the total brightness (Tonomura, 1984).

A schematic of the general design for a multistage electrodes CFEG is reported in Fig. 2.3. On the same figure, the evolution of the source size

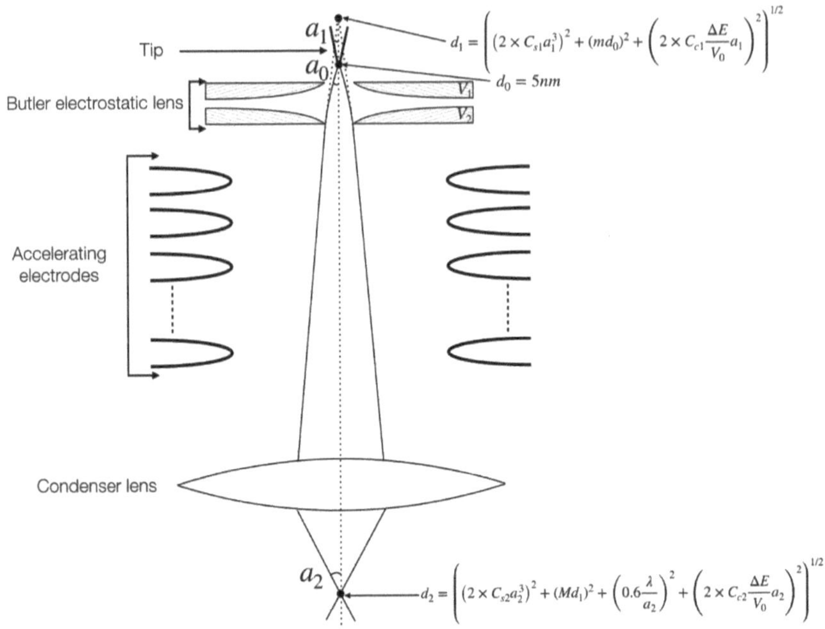

Fig. 2.3 Outline of a multistage acceleration CFEG showing the evolution of the source size from the virtual source size in the emission area to the formation of the focused electron probe in the sample plane. *m* and *M* are respectively the linear magnifications of the Butler lens and the condenser lens. C_{s1}, C_{c1} the spherical and chromatic aberrations coefficients of the Butler lens. C_{s2}, C_{c2} the spherical and chromatic aberrations coefficients of the condenser lens.

image (real or virtual) with respect to aberrations, magnification and diffraction is also indicated for the multistage acceleration gun followed by one condenser lens. To simplify the mathematical expressions all terms coming from the Lagrange–Helmoltz paraxial invariant used to express angular magnification, such as in Eq. (2.7), are hidden in the aberration coefficients. This principle is the same for all multistage gun design whatever the acceleration voltage (from 100 kV to 1 MeV) (Kawazaki, 2010).

In the 125 kV multistage CFEG first implemented in 1984 at HCRL, the distance between the emitter and ground electrode in this multistage electrode gun was approximately 10 times bigger than in the Butler lens gun. This distance reached 1.5 m in the 1 MeV CFEG also created by Hitachi using the same design reported in Fig. 2.3 (Kawasaki et al., 2000).

As already discussed, the main challenge of CFEG technology is to generate stable emission current. Indeed, long-term emission current drift as well as high frequency current fluctuations are always observed in CFEG.

But these intensities can be manageable for practical applications if the beam is generated under ultra high vacuum (UHV). Using standard UHV technology, state of the art CFEG are operated under ~1.10^{-9} Pa. By using this vacuum level, 10 μA beam can be extracted from a *W* [310] cathode and this value will decrease with a ~10%/h rate with ~1% of high frequency current fluctuations (Hawkes, 2009).

Implementing UHV in a multistage CFEG is more challenging than in standard Butler lens CFEG. Indeed, pumping system is electronically referenced relatively to the earth potential and UHV is needed close to the tip area which is then further away from the earth electrode. This distance will increase with the high voltage value as the number of intermediate electrodes will increase as well (Kasuya et al., 2014). To reach 10^{-9} Pa range close to the cathode in a multistage CFEG, apart from selecting an optimal vacuum system, we have to carefully design all the internal electrodes (material choice, shape, ...). The final accelerator tube used to install the electrodes and applied high voltage is a complex ceramic-to-metal brazed assembly which have to be baked out under high temperature (~300–350 °C). Furthermore, special attention should be paid to high voltage conditioning necessary to obtain a proper vacuum level even when high voltage is applied to all electrodes of the multistage CFEG and the beam is fully extracted.

2.3.2 The HF2000 column

All the developments performed in HCRL on high-voltage FEG technologies were then transferred to Hitachi's Naka Works (Kamino and Tomita, 1983). With the aim of enabling new applications in electron holography and nanoanalysis using high brightness STEM, Hitachi launched a project to develop a 200 keV FE-TEM model in 1986. Scientists moved from HCRL to Naka to work on developing this mass-production 200 kV FEG called HF2000 (Isakozawa et al., 1989). This model, described in Fig. 2.4, was one of the world wide top instrument at the end of the 80's. Its nano-probe illumination performance with sufficient beam current (several hundred nA), smaller (<0.4 eV) energy spread, together with the generation of optimum beam spatial coherency in TEM illumination, enabled many applications that were previously impossible (Hashimoto et al., 1991).

Fig. 2.4A reports a schematic of the HF2000 column. This schematic is associated to the picture of the instrument reported in (Fig. 2.4B). The column is equipped with a 200 keV multistage CFEG, which will be described in detail in the following, associated with a double condenser illumination

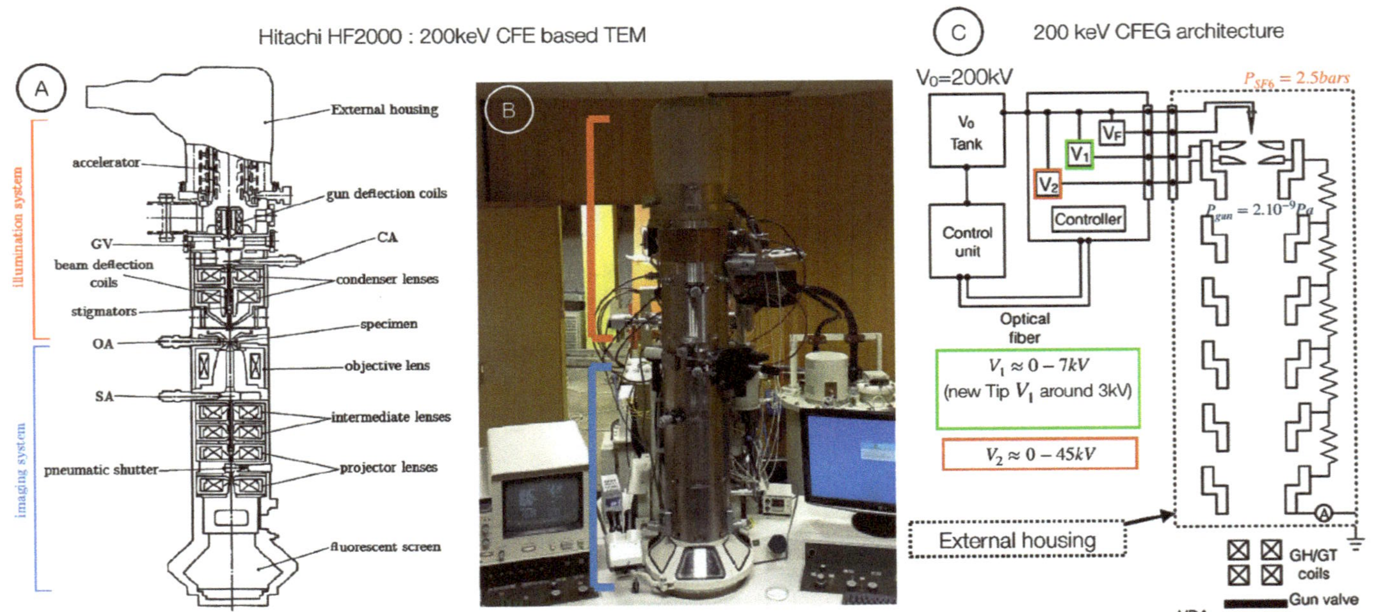

Fig. 2.4 (A) Outline of the Hitachi Field-Emission 200 keV TEM: the HF2000. (B) Picture of one conventional HF2000 FE-TEM used at CEMES. (C) Electric configuration of the HF2000 multistage acceleration CFEG.

system (red lines) and a Ruska type objective lens with a 4.5 mm pole piece gap. The sample is inserted in the center of this pole piece gap located at the maximum of the axial magnetic field of the objective lens, as well as to the eucentric position of a four axis goniometer (four movements: three translations x, y in the sample plane, z along the electron beam, and one tilt α around the sample holder rod).

The imaging system consists of two intermediate lenses and two projector lenses. Four deflection systems are used in the HF2000. Each deflection system consists of a pair of magnetic dipoles (one upper dipole and one lower dipole) for each direction perpendicular to the beam (x and y). Each pair of dipoles produces either a pure shift of the beam parallel to the optic axis (called "Horizontal" in the instrument), or a pure tilt relatively to a specific optical plane which can be adjusted. Gun horizontal X, Y, and gun tilt X, Y (GH(X,Y);GT(X,Y)) are located after the accelerator tube before the gun valve (GV) located above the first condenser lens. Beam horizontal X, Y, and Beam tilt X, Y (BH(X,Y);BT(X,Y)) are located between condenser 1 and 2. STEM scan coils are located in the same optical area as BH,BT coils. Finally image/ diffraction shift can be adjusted using intermediate and projector alignment coils (IA(X,Y),PA(X,Y))) located in the imaging system.

The aperture system is classic and consists of one condenser aperture (CA) located after the gun valve directly below the gun and before the condenser 1 lens. This aperture is used to adjust the beam angle in TEM illumination mode corresponding to the Fig. (1.2A) case (called ZOOM mode). This mode is obtained when condenser 2 is inactive. Another illumination aperture located between condenser 1 and condenser 2 called STEM aperture (not represented in Fig. 2.4A) is used when analysis illumination mode is set, corresponding to the Fig. (1.2B) case (called ANA mode). This mode is obtained with a specific condenser 2 excitation.

Finally a contrast and a selected area apertures (OA and SA) are located respectively in the objective lens focal and image planes. Electron biprism can be also installed in a dedicated port located mechanically close to the SA aperture, but in an optical plane located slightly above the objective lens image plane. Unfortunately, a mechanical interference between SA aperture and biprism holder prevents to use both system at the same time (unlike the I2TEM biprism systems which are more moderns).

The detection system can include STEM detectors such as BF, ADF or HAADF as well as camera system incorporated below the usual fluorescent screen.

2.3.2.1 Focus on the HF2000 200 keV multistage CFEG

Fig. 2.4C reports the electric configuration of the HF2000 multistage CFEG. The V_0 voltage is generated thanks to a Cockcroft–Walton electronic bridge located inside a dedicated tank, filled with mineral oil, called V_0 tank (Cockcroft and Walton, 1932). The voltage is sent to a second tank called "FE-tank" filled with SF6 gaz which is used to control the Butler lens CFEG located above the accelerating tube. This lens can be adjusted by two voltages V_1 (the extraction) and V_2 called gun lens voltage. Such as for any Butler lenses, V_1 is used to set the FE emission current and V_2 can be used to adjust the Butler lens strength. Because the V_1 value will evolve during the tip lifetime, the V_2 voltage will have to follow this change to keep the optical configuration chosen by the user. That's why the Butler lens strength is set through the ratio $R = V_2/V_1$ calculated between the two voltages. Fig. 2.5 reports two axial rays trajectory calculated by SIMION (Dahl, 2000).

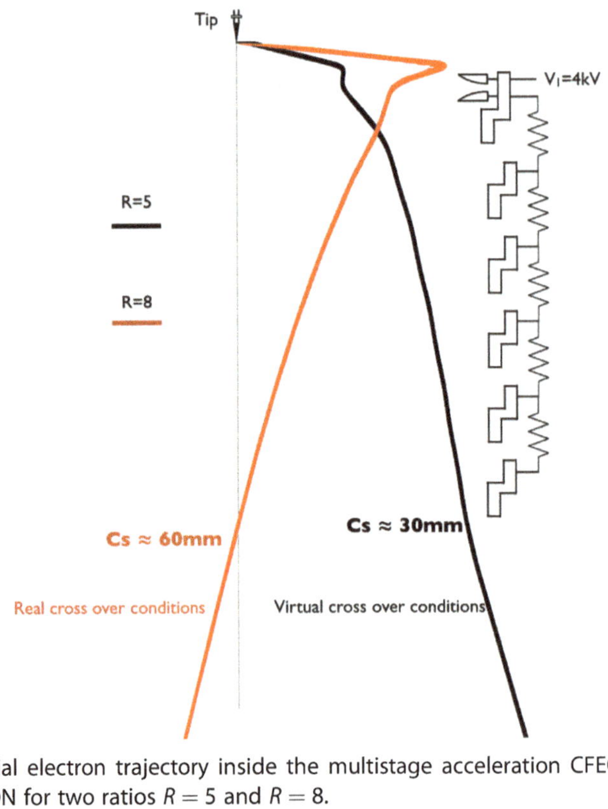

Fig. 2.5 Axial electron trajectory inside the multistage acceleration CFEG computed using SIMION for two ratios $R = 5$ and $R = 8$.

The two rays start from the tip and are focused by the Butler lens using two different ratios $R = 5$ (black trajectory) and $R = 8$ (red trajectory).

Thanks to these simulations we can observe the focalization action of the gun lens going from virtual cross over configuration, obtained using ratios $R < 7$ (black trajectory), to real cross over mode set using ratios $R > 7$ (red trajectory). Virtual cross over configuration are used when high demagnification of the source by the condenser lens is needed. In this mode the aperture aberration coefficient is optimum around $C_s \sim 30\,\text{mm}$. This mode is set for applications requiring an optimum brightness such as high resolution STEM, nanoanalysis, and electron holography. On the other hand, the real cross over configuration allowed to collect more emitted electrons in the final probe at the expense of a smaller condenser demagnification power and a higher gun lens aperture aberration coefficient twice the one calculated in virtual cross over. This mode is set for applications requiring an optimum beam intensity such as conventional TEM imaging. Inside the FE tank, we also find the electronic controlling the flash cleaning of the tip V_F. The three voltages V_1, V_2 and V_F are all referenced to V_0 voltage as reported in Fig. 2.4C. They are injected into the gun area by a second high-voltage connection that runs from the FE tank on one side to a specialized high-voltage plug on the other, known as the external housing or gun housing. Fig. 2.6 shows how the real system works. The housing, made of insulating material, is installed around the gun. The inner parts of the housing are

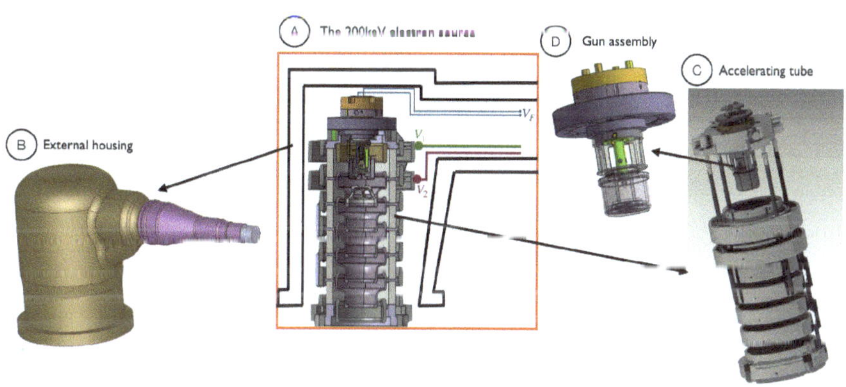

Fig. 2.6 CAD drawings of the HF2000 multistage acceleration CFEG showing—(A, B) the external gun housing attached to the FE-tank on one side (not shown) and installed around the complete electron source on the other side. The electric configuration inside the housing is outlined in (A). (C) The whole electron source is divided into two sections in this technical illustration: The accelerating tube and the gun assembly shown in (D).

springs connected to the 4 voltages signals (V_0, V_1, V_2 and V_F) coming from the high voltage cable. These springs simply touch the gun external electrodes when the housing is plugged around the gun. V_F voltage is applied between the two tip contacts, V_1 to the first large external electrode, V_2 to the second one, and finally the total voltage is equally distributed between the five remaining electrodes from V_2, referenced at V_0, to the ground electrode thanks to five resistors connected between these electrodes. Each of the gun's external electrodes is then connected to the inner electrodes via metallic contacts that are insulated by ceramic components. All this complex structure is sealed using metal-to-ceramic brazing. Between the housing and the gun structure, we inject 2.5 bars of pressurized SF6 gaz to insulate the high voltage applied to the external electrodes (Koch, 2003). The inner part of the gun is evacuated under $\sim 1.10^{-9}$ Pa UHV.

The gun can be separated in two parts: the gun assembly and the accelerating tube. As already described, the accelerating tube is a metal-to-ceramic assembly allowing to connect external electrodes to the internal cylindrical electrodes of the Butler lens and the accelerator. The accelerating tube includes two UHV compatible mechanical ports on each side of the cylinder. On the top side a DN-CF63 flange is used to connect the gun assembly and a DN-CF100 flange on the bottom side is used to connect the gun to the column.

As described in Fig. 2.7, the gun assembly is also separated in two parts: the tip holder and the extraction holder. The tip holder is a DN-CF40

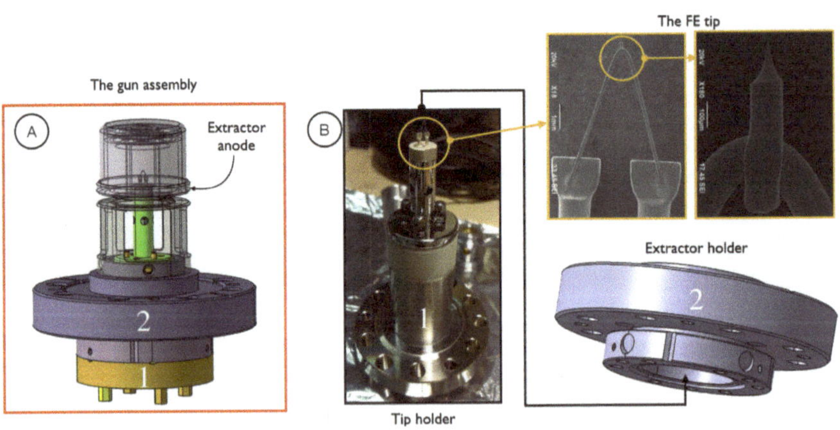

Fig. 2.7 Enlargement of the gun assembly area. (A) General CAD drawing showing the tip holder (1) and the extractor holder (2) used to fix the extractor anode. (B) Picture of the HF2000 tip holder. In the *yellow squares* are reported the SEM images of the FE tip. As shown in (B), the tip holder will be connected to the extractor holder through a DN-CF40 connector.

metal-to-ceramic assembly used to insulate the V_1 voltage applied between the tip and the extracting anode. The FE tip is eventually attached to the tip holder, as shown in Fig. 2.7B, via a metallic cylinder that also connects the V_F voltage to the tip's two pins. The tip holder is finally inserted in the DN-CF40 UHV port of the extraction holder. This port is coupled to a second UHV DN-CF63 port used to fasten the gun assembly on the accelerating tube. A below connects the two flanges, allowing the tip holder to be moved relative to the extractor anode, which is attached to the extractor holder assembly (see Fig. 2.7A). The total range of these two translation movements of the tip in the *x* and *y* directions is 1 mm. They can be driven from the outside of the housing even when the beam is started and accelerated to 200 keV. Carrying out this tip mechanical alignment is possible thanks to four ceramics bars used to actuate 8 mechanical wheels (4 in the ground area and 4 in the high voltage zone) which will then push the tip holder along the chosen direction. These bars can be seen in Fig. 2.6C.

2.4. Improving brightness using new carbon nanocone cathode

My first hardware development, which was aimed at increasing the brightness of the HF2000 electron source, was focused on the cathode rather than the electrostatic optic that focused the electron probe. These primary works aim to see how CFEG properties could be improved thanks to new cathode materials. Many kind of cathode materials have been used as emitter inside a CFEG, but the majority of these works were concentrated on the study of their emission properties using a simple point projection microscope *i.e.* the cathode, the extracting anode and a detector located after the extractor (de Jonge et al., 2002; Nagaoka et al., 1998; Singh et al., 2018; Xu and Huq, 2005). The distance between the extractor and the detector is carefully chosen to have enough magnification enabling the visualization of emission sites on the emitter surface. All these works are very interesting but none of these cathodes were finally used as emitter in a real FE-TEM, even when emission properties seemed very promising. Among all these studies, I was very impressed by the emission properties of carbon nanotubes obtained in the pioneering works of de Jonge et al. (de Jonge, 2009; de Jonge et al., 2002, 2003, 2005). In partnership with Marc Monthioux, a carbon material specialist at CEMES, I chose to pursue that approach, trying to adapt a new carbon nano-cone cathode as a FE TEM emitter (Houdellier et al., 2012, 2015a; Mamishin et al., 2017). The first results were very promising, and a new collaboration around this development was then initiated with HHT slightly after the installation of the I2TEM. Up to now, no Hitachi product proposed an FE-TEM equipped with this new cathode. Indeed, thanks to this very productive collaboration, we quickly understood that the main strength of this cathode, arising from its very small apex size, was also it's major weakness when used in combination with a Butler type CFEG optic to focus the electron probe. The important results of this study will be discussed in the next section, and a possible approach for adapting the CFEG optic to this type of cathode will be investigated in the third chapter, which will be dedicated to my future research project.

2.4.1 Characterization of carbon nanocone field emission cathode

The carbon nanotube (CNT) cathode studied by de Jonge et al. was the ideal candidate since it exhibits very good stability properties of the emitted current as well as high brightness (de Jonge, 2009; de Jonge et al., 2002, 2005). But on the other hand the practical use of CNT cathodes remains strongly limited due to many technical challenges. Indeed, the CNT needs to be correctly mounted onto a filament which includes the laborious task of selecting a CNT which is suitably straight, aligning the selected CNT in the right direction for emission, and fixing the CNT to insure a good electrical contact. Reproducibility is poor due to the difficulty of the mounting operation and variation in field-emission characteristics and conductivity of the CNTs are often observed (de Jonge et al., 2003). Growing a single, well-aligned CNT at the right place is also tricky. A review of the various procedures to implement such a delicate mounting operation can be found in the literature (Ribaya et al., 2008). Furthermore, nanotubes are long and thin making them prone to vibrate, a feature which is able to provoke their breakage during emission and a dramatic degradation of the emission brightness (Saito et al., 2005).

Few years ago, the CEMES carbon material group, around the research activity of Marc Monthioux, developed a new kind of carbon structure in the form of a carbon nanotube-supported pyrolytic cone, that we will called carbon nano-cone or CcNT from now (Allouche and Monthioux, 2005; Allouche et al., 2003; Monthioux et al., 2006). They are prepared by a so-called "time-of-flight chemical vapor deposition" process which consists in depositing pyrolytic carbon onto previously grown CNT used as substrates. It can be precisely described as a carbon-nanotube-supported pyrolytic carbon deposit which includes a short microfiber segment with rough surface and two opposed cones with smooth surface which terminate the short fiber segment ends as observed in the Fig. 2.8A.

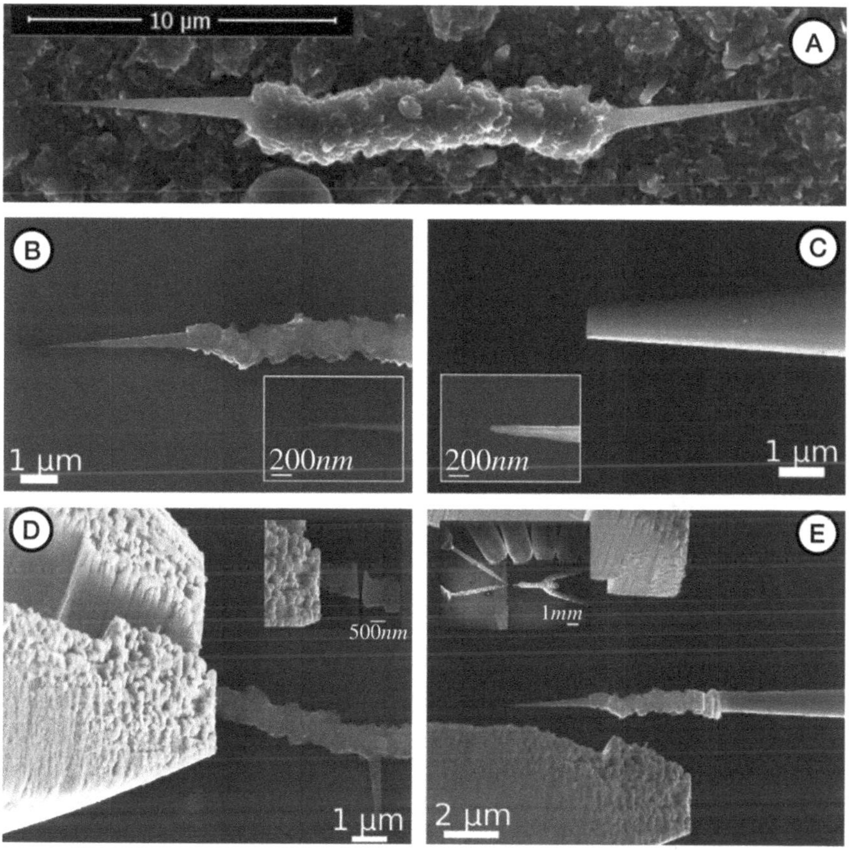

Fig. 2.8 (A) Morphology of a CcNT with a fiber segment (B) CcNT selected to be used for the assembly procedure (similar to A). The inset shows the apex radius of the carbon cone, measured below 10 nm. (C) View of the tungsten tip after being truncated using the FIB. The inset shows the original apex of the tip. (D) The clamp of the carbon cone made with the tweezers using 20 degree tilted stage enables a protection of the tip during electron or ion imaging and ablation. The inset shows the next step when the carbon microfiber is cut using the FIB. (E) Final tip obtained after welding the carbon microfiber to the tungsten tip. The inset displays the whole setup inside the dual beam (tweezers on the *left*, the tungsten support on the *right* and the gas injection nozzles on *top*). *From Houdellier, F., Masseboeuf, A., Monthioux, M., Hÿtch, M.J., 2012. New carbon cone nanotip for use in a highly coherent cold field emission electron microscope. Carbon 50 (5), 2037–2044.*

Installing the CcNT as an emitter inside a FE-TEM first requires the mounting of this cone on the standard filament support. The mounting procedure is performed inside a dual beam FIB using a tungsten support (Yao, 2007). Hence, a regular tungsten tip needs first to be prepared by electrochemically etching of a polycrystalline tungsten wire in a *KOH* solution

(Ibe et al., 1990). The tip is then micro-welded on a regular V-shape TEM tungsten filament. The resulting tungsten support is shown in the inset of Fig. 2.8E (Houdellier et al., 2012).

A suitable carbon cone is then selected using the SEM imaging, according to the proper alignment of the cone axis with respect to the microfiber axis (see Fig. 2.8B). It is then lifted from the surface using micro-tweezers after cutting it from its base using the Ga ion beam (see Fig. 2.8D). The apex of the tungsten tip has been preliminary prepared by cutting it to generate a flat surface once again using the Ga ion beam (see Fig. 2.8C). The cut face of the carbon microfiber segment bearing the cone is finally welded by W ion beam induced-deposition (from a W precursor, namely $W(CO)_6$) on the flat tungsten surface as reported on Fig. 2.8E. The final support, which could be installed inside a standard HF2000 CFEG, is now obtained displaying an apex radius below 5 nm corresponding to the diameter of the multi-wall CNT originally used as substrate to prepare the CcNT. A particular attention has been paid to protect the carbon cone from parasitic deposition or ion irradiation of the cone by tilting the FIB stage by 20 degree as shown in Fig. 2.8D. This allows the cone to be protected from the ion beam using tweezers throughout the procedure (Houdellier et al., 2012).

I will quickly detail the crucial field emission properties of these cones, namely their exit work function and the angular current density of the extracted electrons, before moving on to the characterization of the electron probe generated by a CcNT emitter inside a FE TEM. They were studied first thanks to *in situ* field emission TEM experiments and in a second time inside a point projection microscope developed in the laboratory (an UHV bench fitted with the CcNT emitter + extractor and a fluorescent screen with Faraday cup located few cm ahead of the extractor). All the technical details can be found in the literature (de Knoop et al., 2014, 2015; Houdellier et al., 2015a). The *in situ* experiments were carried out using a Nanofactory holder (double tilt, with field emission, STM and nano-indentation capabilities) inside the SACTEM. The mounting procedure of the CcNT on the nanofactory holder is strictly identical to the one previously described (de Knoop et al., 2014). *In situ* biasing experiments were carried out inside the TEM during the observation (de Knoop et al., 2015). For each extraction voltage V_1, one electron hologram was acquired around the tip apex. Electric field strength around the tip is then extracted from the hologram phase shift relatively to an area taken far from the tip. Additionally emission current I_e as a function of the extraction voltage V_1 has been plotted. $I_e(V_1)$ curve was then fitted following the Fowler–Nordheim emission law given by Eq. (2.4) using F_{loc} determined by the holography experiments. It is worth noting that extracting F_{loc} from an electron hologram is not straightforward and needs comparison with simulations (de Knoop et al., 2014). By combining FEM simulations of the electric field around the apex and electron phase shift calculation using Aharonov–Bohm formula, it was possible to determine a theoretical hologram phase shift. This value was systematically compared to the experimental one. Fitting the $I_e(V_1)$ curve using Fowler–Nordheim law allowed us to determine the exit work function $\Phi = 4.8\,\text{eV}$ of the CcNT emitter (Houdellier et al., 2015a). This value was consistent with the measurements performed on CNT emitter in the literature (de Jonge et al., 2002). Finally, using the point projection microscope, an angular current density in the range of [10–13] $\mu\text{A}\,\text{Sr}^{-1}$ has been determined (Mamishin et al., 2017). This value is also consistent with standard angular current density encountered in CFE using *W* emitter (Hawkes, 2009). This small value compare to Schottky emitter (SFEG) is actually one important property of CFEG. Indeed, combining small angular current density together with high brightness allowed CFEG to generate coherent beam with low energy spread due to the minimization of the Boersch effect. This very important property remains then unchanged using CcNT emitter. These numbers compared to the one determined for W based CFEG and ZrO/W SFEG are reported in the Fig. 2.9.

The emitter has been finally installed inside two instruments: the HF2000 in CEMES and an HD2700 in HHT factory (Houdellier et al., 2012; Mamishin et al., 2017). The HD2700 is a 200kV dedicated STEM described in Fig. 2.11A. The gun of these two instruments are strictly identical. However they were complementary to study the imaging capabilities of this emitter using the two illumination setups described in Fig. (1.2). Indeed, our HF2000 didn't have the STEM capabilities at that time. As a result, the imaging capabilities of CcNT were investigated using the HF2000 illumination system's TEM optical configuration, corresponding to a sample plane conjugated with the illumination pupil plane. The HD2700 works using STEM optical configuration where the sample plane is conjugated with the image plane of the illumination system allowing to focus a small probe. Furthermore, regarding potential application market, HHT was more interested in the HD2700 STEM results than the coherent TEM capabilities, such as holography, measured using the HF2000 in CEMES. TEM and STEM capabilities will be described in the next paragraph.

To conclude this part dedicated to field emission properties of CcNT, it is important to look at the emission current stability. They were studied

	CcNT CFEG	W [310] CFEG	ZrOW[100] SFEG
Exit work function (eV)	4.8	4.25	2.92
Angular current density J (μA/Sr)	10 - 13	~10	~100
Energy Spread (eV)	0.34	~0.3	0.7-1
Brightness (A/cm².Sr.V)	10^5	10^4	10^3
Virtual source size (nm)	1	~5	~30
Emission decay	<20% of probe current in 9 hours	<20% of probe current in 2 hours	No decay
Temperature (K)	300	300	1700
Vacuum Pressure (Pa)	10^{-9}	10^{-9}	10^{-6}
Probe current noise	<2%	<2%	<1%

Fig. 2.9 Table summarizing the main source properties. *All the numbers reported for standard W [310] CFE source and ZrO W [100] SFE source have been gathered from Hawkes, P.W. (Ed.), 2009. Cold Field Emission and the Scanning Transmission Electron Microscope. Advances in Imaging and Electron Physics, vol. 159 and Swanson, L., Schwind, G., 2008. Review of ZrO/W Schottky cathode. In: Handbook of Charged Particle Optics, second ed. CRC Press, pp. 1–28.*

using both instruments by measuring the probe current stability with a Faraday cup located at the sample plane. Gun optical conditions were the same for the two instruments and the results obtained with CcNT emitter were then identicals. These measurements were finally compared to the datas measured using *W* emitter (Mamishin et al., 2017). To have a fair comparison, the chosen datas were the best one obtained by HHT with their *W* emitter measured under optimum conditions (new tip, best vacuum level, optimum optical configuration). The tip was set at $V_0 = -200\,\text{kV}$. The distance between the tip and the anode is adjusted in a range of about 4–7 mm. The ratio $R = V_2/V_1$ between the focusing anode and the extracting anode can be adjusted between 2 and 13 on both instrument. Fig. 2.10 reports the stability data of a 2 nA probe at 200 kV measured over 9 h without any flash cleaning of the tip during the measurement (Mamishin et al., 2017). Fig. 2.10A corresponds to the CcNT emitter with a total emission current set at 5.7 μA, an extraction voltage $V_1 = 1.3\,\text{kV}$ and a ratio of 13, while Fig. 2.10B refers to the standard *W* [310] based CFE source with an emission current of 10 μA for an extraction voltage of 2.9 kV and a ratio of 4.5. The source's vacuum system was the same for the two instruments

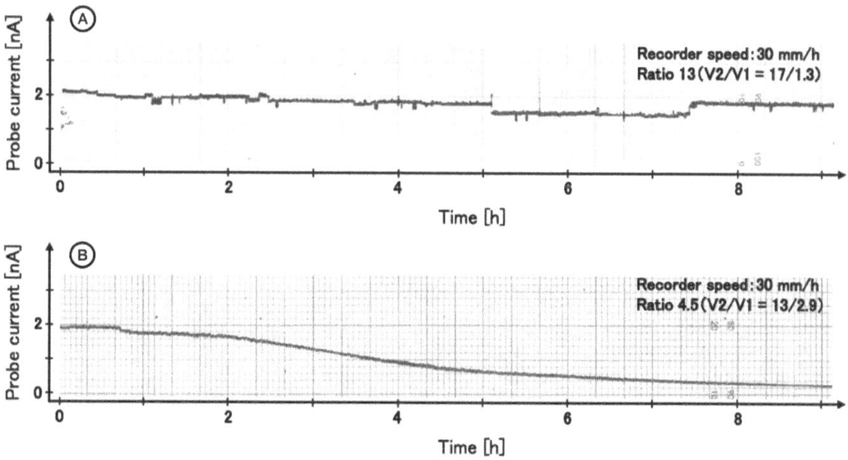

Fig. 2.10 Nine hours stability measurements, performed on a Hitachi HD2700 dedicated STEM, of a 2 nA probe with 200 keV of energy. Comparison between CcNT CFE source (A) and *W* [310] CFE source (B). Both measurements were started after a flash-cleaning of the tips. *From Mamishin, S., Kubo, Y., Cours, R., Monthioux, M., Houdellier, F., 2017. 200 keV cold field emission source using carbon cone nanotip: application to scanning transmission electron microscopy. Ultramicroscopy 182, 303–307.*

and for the two measurements (CcNT and *W*), fitted with two ion pumps and one non-evaporable getter (NEG) pump reaching a vacuum level below 2.10^{-9} Pa. As Fig. 2.10 clearly demonstrates, the long-term stability is strongly improved with the CcNT used as FE cathode. Indeed, using state of the art *W* [310] emitter installed in an optimized environment, the probe current decreases by a factor of 50% after 4 h and by almost 90% after 8 h while under the same vacuum condition the CcNT CFE source shows no decrease of the probe current over the full measurement time range *i.e.* 9 h. The presence of tiny peaks can be seen, although they have no bearing on STEM or TEM imaging. The root mean square value of the probe current noise is however not improved and is measured around 1%, which is in the order of magnitude of standard CFEG technology using *W* emitter (Hawkes, 2009).

2.4.2 Application to S/TEM

Imaging properties were first studied using the TEM configuration inside the HF2000. In particular, comparison between coherent properties of the illumination system, fitted either with a CcNT or a standard *W* emitter, were carried out looking at condenser aperture Fresnel fringes (Houdellier et al., 2012). Electron holograms were also recorded under the same conditions using the two emitters. However, the use of off-axis electron holography to unambiguously compare the coherence properties of the two sources is not straightforward due to the contribution of various perturbations described in Section 1.1.3.3 that could induce a contrast damping. Comparison of Fresnel fringes was then preferred to carry out this comparison. By combining Fresnel contrast measurement with the probe current and the angular current density measurements, determined in the previous experiments, we have measured a reduced brightness of $B_r = 1.6 \times 10^5$ A cm^{-2} Sr^{-1}V^{-1}. This value is higher than the state-of-the-art reduced brightness obtained using W based CFEG on commercial instrument which is in the range of few 10^4 A cm^{-2} Sr^{-1} V^{-1}. Finally, using Eq. (1.38), we have also determined the virtual source $d_v = 1$ nm which is five times smaller than the one measured with the W emitter (Mamishin et al., 2017). This factor 5 is the same for the brightness as the angular current density remains identical using the two emitters. All the details concerning these measurements can be found in the literature (Houdellier et al., 2012, 2015a; Mamishin et al., 2017). These first encouraging results seemed to show us that CcNT could be an interesting candidate as a new emitter to improve coherent TEM capabilities. However, when considering the imaging capabilities of CcNT in a STEM setup, the situation becomes slightly more complicated, yielding results that are far more mixed in terms of prospective applications.

Compared to a conventional W [310] tip the CcNT apex is smaller, which is its main strength regarding previous brightness measurements. But this will have important consequences on the Butler lens properties. Indeed, the extraction voltage V_1 required for a specific probe current value will then have to be smaller. This is the case in Fig. 2.10 where 2 nA is obtained using $V_1 = 1.3$ kV of extraction voltage using the CcNT cathode while we have to apply $V_1 = 2.9$ kV using a new W cathode. This difference will change the electron optical configuration of the gun because it changes the strength of the electrostatic Butler lens. This effect was simulated and reported in Fig. 2.12B and C showing the evolution of axial electrons trajectories calculated using SIMION considering 5 ratio values under the same emission conditions as those described in Fig. 2.11.

We can see that, whatever the ratio value, the 200 kV gun cross-over remains real when the CcNT is used as a cathode. As already introduced in Fig. 2.5, the same trajectories obtained using W cathode (reported in Fig. 2.12C) show a gun cross over which can be switched from a virtual position to a real one for ratios taken between 6 and 8. This effect is clearly observed looking at the ratio 2, which corresponds to the condition with the

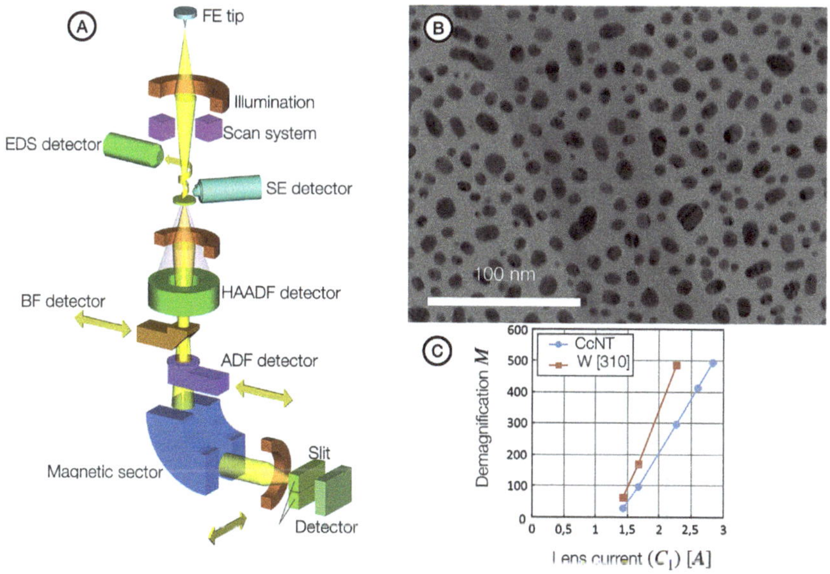

Fig. 2.11 (A) Configuration of the dedicated STEM Hitachi HD-2700. (C) BF-STEM image of gold nanoparticles obtained with the HD-2700 fitted with CcNT CFE source set to a ratio $R=13$. Emission conditions were the same as the one reported in Fig. 2.10A. (D) Comparison of C_1 lens demagnification power of the gun cross-over obtained with a CcNT and a *W* [310] CFE source. *From Mamishin, S., Kubo, Y., Cours, R., Monthioux, M., Houdellier, F., 2017. 200 keV cold field emission source using carbon cone nanotip: application to scanning transmission electron microscopy. Ultramicroscopy 182, 303–307.*

minimum strength of the Butler lens. Even using this condition the cross-over remains real and located below the accelerator when the CcNT cathode is used, while it should be virtual in a conventional CFEG. This unwanted effect, related to the small apex size of the CcNT, has a direct consequence on the STEM properties. Indeed, for STEM imaging the virtual CO condition is the most desirable in order to maximize the demagnification power *M* of the illumination lenses which is strongly related to the distance a_1 between the gun cross-over and the condenser lenses, as explained in Fig. 2.12A. Furthermore, as already mentioned previously, real cross-over condition is not optically optimum due to higher contribution of spherical and chromatic aberration coefficients of the Butler lens which are respectively $C_s \approx 60$ mm and $C_c \approx 12$ mm (instead of 30 mm and 10 mm in virtual CO condition) (Murakoshi et al., 1992). Of course, as the Butler lens demagnification M_G grows, the aberrations effect reduces, but if the CO stays real, the characteristics never reach those achieved in the virtual CO state.

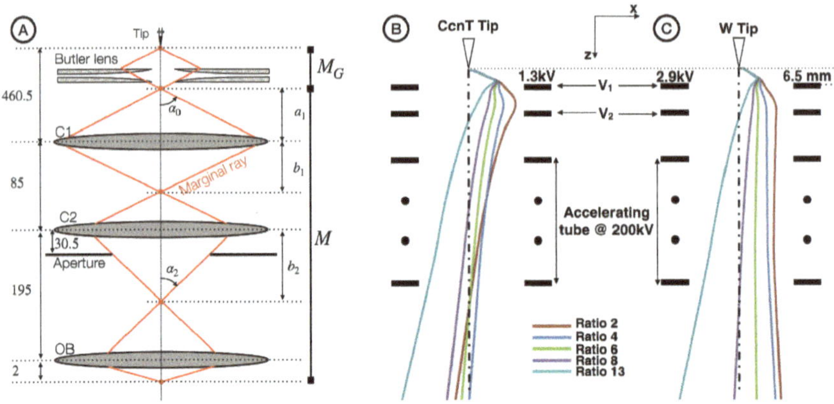

Fig. 2.12 (A) General optical outline showing the source demagnification principle combining the effect of the gun electrostatic lens demagnification M_G and the one coming from the total action of the illumination lenses M. (B, C) Axial electron trajectories inside the HD-2700 source calculated with SIMION 8.1 (Dahl, 2000). Trajectories reported in B correspond to the evolution as a function of the ratio, of one axial ray extracted from the CcNT CFE source set to 5.7 μA of emission current corresponding to $V_1 = 1.3$ kV. For each ratio, the initial conditions are the same. Trajectories reported in C correspond to the evolution as a function of the ratio, of one axial ray extracted from the W [310] cathode set to 10 μA of emission current corresponding to $V_1 = 2.9$ kV. To improve the visibility the trajectories are enlarged in the x direction by a factor 500 and superimposed to a sketch of the source. *From Mamishin, S., Kubo, Y., Cours, R., Monthioux, M., Houdellier, F., 2017. 200 keV cold field emission source using carbon cone nanotip: application to scanning transmission electron microscopy. Ultramicroscopy 182, 303–307.*

Then, to maximize the illumination demagnification M, the distance a_1 must be as large as possible. Using the CcNT emitter the only solution was to increase the Butler strength to a ratio of 13 (see Fig. 2.12B). It's worth noting that, in addition to the poor aberration properties provided by this condition, it's also not ideal for an EELS application because the cross-over is placed in a zone where the electrons have a low energy, which is favorable to energy spreading due to the Boersch effect. Unfortunately as shown in Fig. 2.11C, even using this ratio condition, the demagnification M does not reach the level typically achieved with a W based CFEG set in virtual CO with a certain C_1 strength. Indeed, considering a maximum C_1 lens current of 3 A, the demagnification remains limited to 400 with a CcNT source and a ratio of 13, while it could reach almost 700 using a W source with a ratio of 4.5 (Mamishin et al., 2017).

The BF-STEM image of gold nano-particles acquired using the CcNT CFE source, set at 200 kV with a ratio of 13, is reported in Fig. 2.11B.

Although the overall STEM image has a good quality, it was impossible to reach the desired atomic resolution due to this lack of demagnification *M*.

2.4.3 Discussion

We have significantly increased the probe current stability of a CFE source by using a CcNT as a field emission cathode inside a 200 kV dedicated FE S/TEM. This makes it a serious challenger to SFE sources in the TEM, STEM, and even SEM markets. Besides its very good stability, the CcNT CFE source preserves all the excellent properties encountered using standard CFE source *i.e.* high brightness together with low current angular density, small virtual source size and low energy spread. Indeed, as compared to a normal W-based CFE source, the reduced brightness has grown by a factor of 5, owing to a virtual source size reduction of nearly the same factor. As a result, the spatial coherence properties of the produced beam in the TEM illumination condition have been improved. These excellent properties are however strongly limited in STEM illumination configuration due to the electron optical configuration of the original Butler electrostatic lens. Combining CcNT cathode with a Butler lens generates a first real cross over inside the accelerator, which is responsible for a lack of demagnification power of the illumination system and a stronger contribution of the gun lens aperture aberration. To take full advantage of the properties provided by the CcNT cathode, a dedicated electron optical configuration will have to be designed to see beyond standard Butler optic in CFE. Indeed, this simple electrostatic optic was used almost since the birth of CFE, and was always improved from time to time but never changed in depth as far as medium acceleration voltages commercial instruments are concerned. Using this CcNT we have now reached the limit of this technology. Hence, in the last chapter dedicated to my future research project, I will discuss possible solutions to overcome this problem. The one based on an pre-accelerated magnetic gun lens introduced long time ago by Troyon et al. is the one which seems practically feasible in combination with a 200 kV accelerator (Kasuya et al., 2014; Troyon, 1980). First theoretical calculations have shown very promising results ! Our future work will move toward that direction. Before I go into the details of this project, I will end up this second chapter with a description of another instrument I have been working on with new CFE hardware. By combining a CFE with an ultrafast Laser we succeed to generate ultrashort coherent electron pulses (few hundred of femtosecond). This new ultrafast CFE source has been installed inside a HF2000 to implement new coherent ultrafast TEM investigations using pump-probe setup. Such as for CcNT CFE source, we will see how this new ultrafast CFE source is also strongly limited by the poor properties of the Butler electrostatic lens. The development of new CFE optic proposed within my future project will then be highly beneficial for both design.

2.5. Generation of ultrashort electron pulses using femtosecond laser pulses exciting a FE cathode

2.5.1 Introduction

The observation of a physical system as it evolves in real time is essential to the understanding of its dynamics. From the beginning, studying the dynamic of samples under excitation was very much part of the I2TEM project as described in Section 1.3.2. Indeed, the I2TEM instrument has been carefully designed to allow a full and an optimum use of cumbersome *in situ* sample holders. Thanks to these holders it is possible to apply an external biasing, traction or even magnetic field excitation on the sample during its observation. Many interesting *in situ* or even *in operando* experiments can be performed within millisecond to hundred of microseconds timescale, which are the timescales compatible with the frame rate of CCD/CMOS cameras. For instance studying the structural dynamics of materials can be carried out using traction experiments looking at the evolution of crystalline defects thanks to diffraction contrasts imaging (BF/DF imaging) (Legros et al., 2008; Mompiou and Caillard, 2014; Mompiou et al., 2012). Also, understanding the fine operation of microelectronic devices at the nanoscale, from single nano-capacitor to more complex and state-of-the-art transistors, is possible thanks to the unique combination of *in situ* stimuli and holography field mapping given by the I2TEM (Einsle et al., 2015; Denneulin et al., 2018). These topics are investigated by many scientists in CEMES, such as Dr. Christophe Gatel, Dr. Frédéric Mompiou, Dr. Marc Legros, etc. who are regularly using the I2TEM, or other *in situ* dedicated instruments (Gatel, 2020; Mompiou, 2015).

However the timescale of many important processes of the dynamics in solids lies typically in the femtosecond to nanosecond range *i.e.* far beyond the reach of the I2TEM, limited by the detector frame rate (Vaz et al., 2008). Indeed, despite all the numerous technical improvements in electron optics, electronics, vacuum, etc. investigations by TEM have been always restricted to systems either static or evolving very slowly compared to the ultrafast dynamic of many phenomenon at the nanoscale. For instance electron microscopes were capable of looking at static atoms but were not fast enough to monitor their displacement on the appropriate timescale *i.e.* picosecond timescale considering the simplest vibration mode of a spherical gold nanoparticle.

Hence, after carrying out, together with HHT, the development, the installation and the final fine adjustments of the I2TEM instrument, I have decided to follow a new instrumental direction together with Dr. Arnaud Arbouet combining our knowledge in ultrafast laser science and cold field emission technology. The idea was to develop a prototype of ultrafast source generating femtosecond coherent electrons pulses allowing to improve considerably the time resolution of a conventional FE-TEM. Indeed combining a bright ultrafast electron source with a well equipped FE-TEM, such as the I2TEM, will open many developments opportunities in ultrafast science. For instance, it will be possible to study the femtosecond dynamic of magnetic systems by time-resolved electron holography, or the vibrational properties of crystals by time-resolved CBED, or nano-optical properties of materials using time-resolved EELS, cathodoluminescence, etc.

The idea of time-resolved TEM is not new. Following the pioneering work of Oleg Bostanjoglo at the Technical University of Berlin, spectacular progress have been made to provide TEM with a constantly improving temporal resolution (Bostanjoglo, 1989, 2002). These developments rely mostly, although not exclusively, on ultrafast lasers and are largely inspired from time-resolved optical spectroscopy techniques (Shah, 1999). Indeed, the generation of femtosecond optical pulses shorter than the characteristic times of material dynamics and the ability to synchronize optical excitation (the pump) and detection (the probe) made possible the acquisition of "snapshots" at regular time intervals Δt during the evolution of a system. These kind of experiments are then called "pump-probe experiments" and are schematically described in Fig. 2.13A. All time-resolved TEM experiments with temporal resolution in the nanosecond range or faster rely on pump-probe experiments involving an optical pulse and a delayed electron pulse as described in Fig. 2.13B (Bostanjoglo et al., 2000; Zewail and Thomas, 2010). The beam from a first pulsed laser is sent to the electron source to trigger the emission of electrons. A second beam from a pulsed laser is sent inside the objective lens of the microscope and focused on the sample. The optical pump pulse brings the sample out of equilibrium and the electron probe pulse, delayed by Δt and synchronized with respect to the excitation, is used to probe the sample during its relaxation. By systematically changing the delay between pump and probe, it is possible to record the dynamical evolution of the sample as the latter goes back to

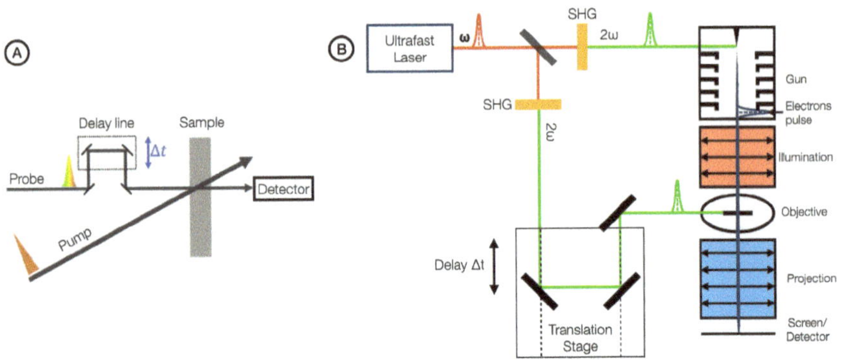

Fig. 2.13 (A) Scheme of the pump-probe optical setup set-up with the pump pulse photoexciting the sample, while the probe pulse is used to monitor the sample response at different times after photoexcitation Δt. (B) Principle of ultrafast pump-probe TEM experiments. The beam from a femtosecond laser source is split in two parts. The first part of the beam is sent to the electron source of the TEM where it triggers the emission of ultrashort electron pulses. Frequency conversion such as Second Harmonic Generation (SHG) is used to improve the emission efficiency. The second part of the beam is sent to the objective lens. It is focused on the sample and brings the sample out of equilibrium, initiating its dynamics. The pump-probe delay Δt is adjusted by controlling the position of a mechanical translation stage.

equilibrium. Therefore, the development of time-resolved TEM requires the modification of two areas of a conventional TEM allowing the injection and the focalization of an ultrafast laser beam: the electron source and the objective lens area (see Fig. 2.13B).

Time-resolved TEM experiments can be performed in two different modes described in Fig. 2.14A and B called respectively Dynamic TEM (DTEM), or sometimes high-speed TEM, and Ultrafast TEM (UTEM) (Arbouet et al., 2018). In the single frame mode DTEM only one electron pulse containing a sufficient number of electrons to get an exploitable signal (image, diffraction pattern, spectrum), is used to probe the sample with a delay Δt with respect to the excitation of the sample by a single pump pulse. By generating a sequence of several electron pulses and displacing the image on the CCD camera with a high speed beam deflector synchronized with the laser, it is possible to acquire multiple frames at different delays Δt with respect to the initiating optical pulse: this mode of operation is called movie-mode DTEM (LaGrange et al., 2011). These two DTEM modes are reported in Fig. 2.14A. In the stroboscopic mode, a very large number (typically $> 10^8$) of excitation-observation cycles are used to accumulate signal. This mode cannot be used to investigate irreversible processes which could be studied only using DTEM. However, it enables TEM measurements

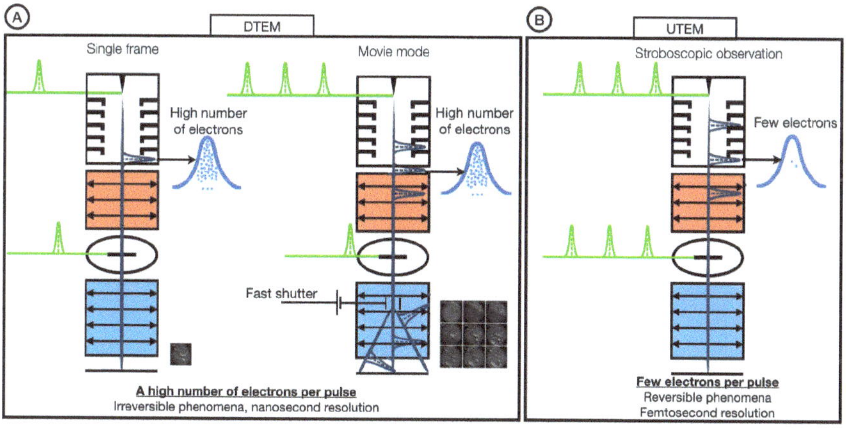

Fig. 2.14 (A) *Left*: single-shot DTEM: a single electron pulse containing a large enough number of electrons (typically $> 10^8$) is used to probe the sample after excitation by one optical pulse. *Right*: movie-mode DTEM: a sequence of several electron pulses is used to probe the evolution of the sample at different delays after excitation by a single optical pulse. (B) Ultrafast TEM in stroboscopic mode: a large number of excitation/detection cycles (typically $> 10^8$) is used to accumulate signal. In the single electron regime, each electron pulse only contains a few electrons therefore minimizing Coulomb repulsion and enabling TEM with sub-picosecond temporal resolution.

with electron pulses containing only a few electrons per pulse. This so-called single-electron regime allows minimizing the Coulomb repulsion inside the electron pulse and brings the temporal resolution into the femtosecond range (Zewail, 2006). Time-resolved TEM in the stroboscopic mode and single electron regime is called Ultrafast TEM (UTEM) due to its better spatio-temporal resolution and is schematically reported in Fig. 2.14B (Zewail and Thomas, 2010).

The first DTEM developed in Berlin by the team of O. Bostanjoglo was later improved in the Lawrence Livermore National Laboratory (Bostanjoglo et al., 2000; Reed et al., 2010). A major breakthrough was achieved in 2005 in the group of A. Zewail at the California Institute of Technology (Adhikari et al., 2017). By using electron pulses containing only a few electrons they succeeded in improving for the first time the spatio-temporal resolution of a DTEM. This first world wide UTEM opened a completely new window on the ultrafast dynamics at the nanoscale (Zewail and Thomas, 2010). However all these developments were performed on brightness limited electron source which are the ultrafast equivalent of thermionic source in conventional TEM. No coherent techniques such as holography are available using this kind of instrument. A schematic outline of these incoherent ultrafast sources is reported in Fig. 2.15A.

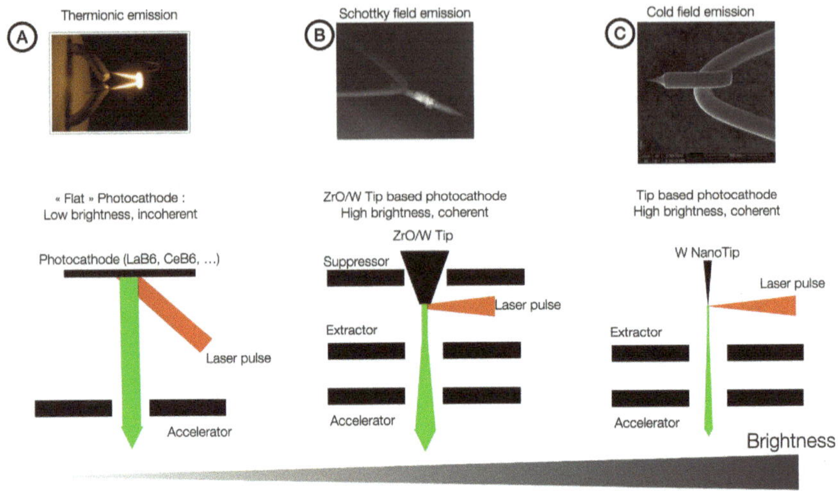

Fig. 2.15 Schematic showing the general strategy used to generate ultrafast electron pulses by combining an ultrafast laser pulse with conventional electron source technologies. (A) Emission from a large photocathode performed by sending a laser on a flat photocathode located inside a conventional thermionic assembly. To apply this strategy to a TEM, the laser pulse is usually sent from the bottom of the gun through the accelerating tube and the Whenelt hole (not represented here). (B) Emission generated from the interaction of a laser pulse with the apex of a ZrO/W emitter located slightly below the suppressor electrode inside a SFEG assembly. (C) Emission generated from the interaction of a laser pulse with the apex of a W nano-emitter inside a CFEG assembly. The electron emission brightness generated by these ultrafast sources increases from left to right such as for conventional source technology.

An ultrafast laser shines a flat photocathode, in general made of LaB_6 crystal. The electrons are extracted by photoemission from the photocathode surface and are then accelerated toward the illumination lenses. Thanks to their ability to extract a high number of electrons in one shot, photocathode sources are well adapted for DTEM, but their limited brightness, linked to their large source size, restricts their capabilities in UTEM mode and prevents the implementation of coherent techniques.

An alternative to the flat photocathodes implemented in UTEMs appeared one decade ago when the emission of electrons from metallic nanotips, similar to the ones used in SFEG or CFEG, could be triggered using a low power, high repetition rate, ultrashort laser beam focused on the tip apex (Arbouet et al., 2012; Hommelhoff et al., 2006a,b; Ropers et al., 2007). Recently, a lot of efforts have been put in the development of UTEMs based on high brightness ultrafast electron sources. The team of C. Ropers in Göttingen university succeeded in modifying a SFEG enabling the generation of bright ultrashort electrons pulses (Bormann et al., 2015; Feist et al., 2015, 2017).

In conventional mode, the emitter is heated at a temperature of about 1800 K. The high temperature of the tip melts a droplet of zirconia coming from the reservoir resulting in the creation of a ZrO overlayer on the (100) front facet of the W tip. This additional layer reduces the work function down to 2.9 eV. In the ultrafast photoemission mode, the temperature of the nanotip is reduced below 1400 K to avoid continuous emission. Electrons are generated with the laser focused at the tip apex by one-photon photoemission from the *W* [100]/ZrO tip as schematically described in Fig. 2.15B. The brightness difference measured between the photocathode and the SFEG ultrafast source were in the same order of magnitude as the one determined between conventional thermionic and SFEG emitter *i.e.* from 10^2 to $10^4\,A\,cm^{-2}\,Sr^{-1}\,V^{-1}$ (Feist et al., 2017). Therefore, our idea was to further improve the brightness of ultrafast source by an order of magnitude using a CFEG technology as illustrated in Fig. 2.15C (Caruso et al., 2017; Houdellier et al., 2018). Indeed conventional SFEG and CFEG brightness are respectively 10^4 and $10^5\,A\,cm^{-2}\,Sr^{-1}\,V^{-1}$. Improving UTEM brightness should enable us to implement time resolved electron holography which was not implemented even using the SFEG UTEM of Göttingen (Houdellier et al., 2019). Finally it worth noting that the use of bright electron source in a DTEM mode is pointless due to the catastrophic brightness deterioration coming from the electrons–electrons interaction inside a DTEM pulse and increasing during the pulse propagation through the instrument. The ultrafast CFEG source described in the next parts will be therefore exclusively used for coherent UTEM applications.

These developments have been implemented during the PhD thesis of Giuseppe Mario Caruso and all the details have been described inside his manuscript (Caruso, 2019).

2.5.2 Development of the first CFEG based UTEM

Similarly to the first implementation of the CcNT cathode, the development of the ultrafast CFEG prototype has been carried out using the HF2000 test bench. Indeed, modifying directly the I2TEM source would have been too much risky. However the new source was designed from the beginning thinking of an hypothetical retrofit onto the HF3300 gun which is, apart from the accelerator size, almost identical to the HF2000 one. Furthermore, the new source was designed using standard [310] oriented monocrystalline *W* emitter. Indeed, the interaction between this material and an ultrafast laser pulse has been intensively studied in the literature especially in the P. Hommelhoff's group (Hommelhoff et al., 2006a,b; Krüger et al., 2011; Schenk et al., 2010). Using a CcNT would have been also too much risky due to a too large number of scientific and technical unanswered questions. Carrying out emission of ultrashort electrons pulses from a W tip is possible by strongly focusing the ultrafast laser beam on the emitter apex with the laser beam arriving perpendicular to the tip. The emission process will depend on the extraction voltage V_1 and on the so-called Keldysh parameter γ describing the ratio between the tunneling time and the optical cycle duration of the laser (Hommelhoff et al., 2006b). This parameter is mainly linked to the laser intensity. The process can be multi-photon photoemission as well as photo-assisted field emission for $\gamma >> 1$ or optical field emission for $\gamma << 1$. Privileging low laser intensity emission process (high γ), we were looking for field-assisted photoemission. In that case, the extraction voltage V_1 should be decreased preventing continuous field emission of electrons from the Fermi level E_F. However thanks to the laser, one or two photons (depending on the photon energy E_{photon} and the emitter exit work function Φ) could be absorbed by the electrons located at the Fermi level to higher energy levels E in the metallic tip *i.e.* $E - E_F > 0$. These electrons could then be emitted by tunneling effect from these energy levels E to the vacuum if V_1, and the associated F_{loc}, remains sufficient to provide a small potential barrier at this energy location (see Fig. 2.1).

Implementing this kind of emission process inside the HF2000 CFEG technology, described in Section 2.4, have requested numerous modifications (Caruso et al., 2017; Houdellier et al., 2018). Before all, the chosen laser system is a commercial instrument based on a compact ultrafast fiber laser, delivering ultrashort (350 fs), high energy (up to 20 μJ) pulses at 1030 nm with a repetition rate f_{rep} tunable between single-shot and 2 MHz. Electron emission is triggered by the femtosecond pulses generated at 515 nm by Second Harmonic Generation (SHG) of the laser output in a beta barium borate (BBO) crystal to increase E_{photon}. The laser beam needs first to be driven from the outside of the gun to the UHV area of the tip and then focused on the tip apex.

Bringing the laser beam in the emitter UHV area requires two major modifications of the electron gun. As a first step, the beam will have to be driven from the outside to the intermediate SF6 area inside the gun housing, previously introduced in Section 2.3.2 and Fig. 2.6. The original Hitachi housing cannot be easily modified to open an optical window. Indeed, the manipulation of the resin material used to insulate the high voltage inside the housing is a too complex technology for us. We have then designed our own gun housing from scratch. This new important element

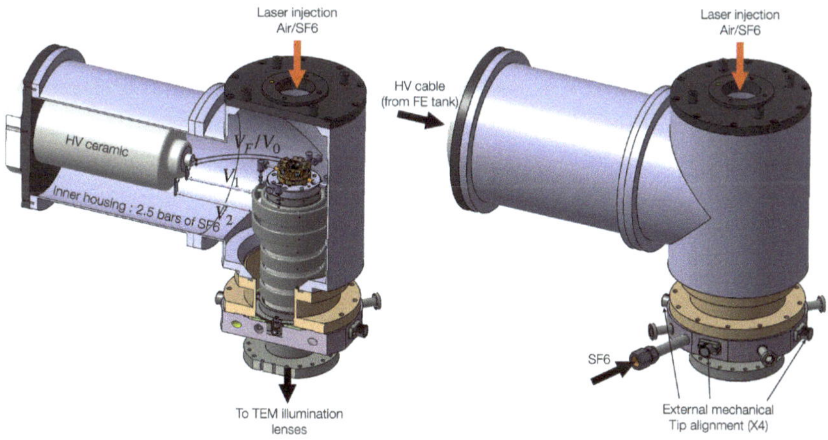

Fig. 2.16 CAD drawing displaying the design of the new 200 kV metallic housing filled with 2.5 bars of SF6 gaz. The electric configuration inside the housing is outlined. The window located at the top of the housing is used to inject the laser pulse above the gun assembly from the outside to SF6 pressured area.

is described in Fig. 2.16. The conception is fully metallic and is constituted by two cylindrical parts. The first cylinder incorporates the high voltage ceramic (HV ceramic in Fig. 2.16). The cable from the FE tank can then be connected to this ceramic on one side, outside the housing. All the signals driven from the tank (V_0, V_F, V_1 and V_2) thanks to the cable could then be separated on the other side of this ceramic located inside the SF6 area. The HV ceramic is attached to the housing with a specific O-ring preventing SF6 leakage. From the ceramic to the external electrodes of the gun, specific insulated wires are installed to bring the V_F and V_0 signals to the emitter, V_1 voltage to the extracting electrode and finally V_2 to the focusing electrode. This first cylinder, attached to the HV cable on one side, is fixed to the second metallic cylinder on the other side. This second metallic part of the housing is installed around the gun. The complete housing is finally attached to the column using a specific flange seen on Fig. 2.16. To bring the laser inside the emitter area we finally choose to inject the beam from the top of the gun which prevents to modify the accelerator (Caruso et al., 2017). The top part of the metallic housing was then separated from the two main cylinders. A distinct metallic plate, with a hole located in the middle, is fixed on the top of the housing. This plate is reported in black in Fig. 2.16. A glass window could then be attached on the top of this hole with another O-ring used to avoid any SF6 leakage. Thanks to this new housing, the FE tank signals could then be safely linked to the external electrodes while maintaining 2.5 bars of insulating SF6 gaz and allowing the laser beam to be injected in this first intermediate area. High voltage stability was first better with the Hitachi original housing made with insulating resin. Due to surface charging occurring on the inner surfaces of our metallic housing, some high-voltage instabilities were observed. Positioning teflon insulating materials (*i.e.* polytetrafluoroethylene: PTFE) between the inner metallic surfaces of the housing and the external electrodes of the gun have prevented us the generation of micro-discharging at the origin of these HV instabilities.

In the second step, the laser beam will have to be injected from the inner housing area to its final location in the vicinity of the emitter maintained under UHV. Injecting the laser pulse in this zone requires the modification of the tip holder presented in Fig. 2.7. Indeed, the laser beam arriving from the top of the gun assembly, only this part has to be modified in the complete gun. During almost 2 years a new tip holder has been designed in collaboration with the Sociéte des Céramiques Techniques (SCT) company located in Tarbes (SCT Company, n.d.). This company is indeed specialist in the fabrication of customized and complex metal-to-ceramic assemblies, a technology not accessible in CEMES. The final tip holder is reported in Fig. 2.17A. Standard alumina has been used to insulate the V_1 potential maintained between the tip holder and the other part of the gun assembly. Through this white alumina, seen in the picture of Fig. 2.17A, one central metallic contact surrounded by three identical contacts are used to apply the V_0 reference voltage, and the voltage difference V_F between the central electrode and the three external ones. Three optical windows have been also inserted between the three contacts. These windows are made of transparent alumina which are finally brazed to the main white alumina. Prior to the fabrication of the new tip holder, the influence of this transparent alumina window on the laser beam properties (pulse duration, polarization mainly) has been studied and did not indicate any detrimental influences. All the brazing connections, *i.e.* main metallic flange to white ceramic, white ceramic to transparent ceramic and finally white ceramic to metallic contacts, have been leak tested after many baking cycles to a maximum temperature of 450 °C.

The laser is then reflected on a plan mirror before being directed to a parabolic one to focus it on the tip apex. Reflective optics have been privileged to minimize the dispersion of the ultrafast pulses. The focal distance of the latter is $f' = 8\,\text{mm}$ which yields a focal spot diameter of $\approx 3\,\mu\text{m}$. This focalization principle, reported in Fig. 2.17D, is basically the same in miniature as a the one used in a solar oven. The mirror are conductors and gold coated avoiding any charging issues. A mirror holder has been

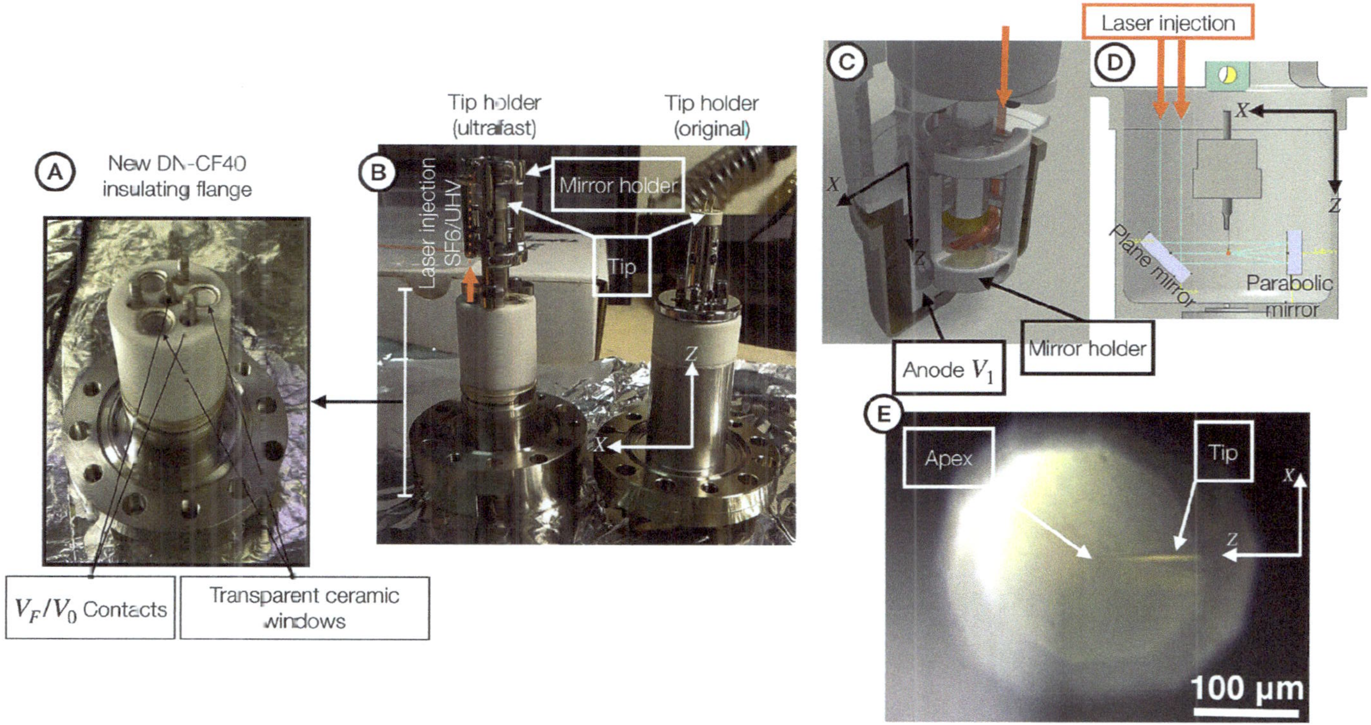

Fig. 2.17 Detailed description of the new ultrafast CFE gun assembly. (A) New tip holder fitted with three transparents ceramic optical windows allowing the injection of the laser beam from the SF6 area inside the housing to the UHV area inside the gun. (B) Comparison of the mounted tip assembly before (*right*) and after (*left*) the modifications required for the focalization of laser beam on the FE tip apex. In particular, the polished mirror holder used to hold the parabolic and the plan mirrors can be seen around the tip. (C) CAD drawing showing the internal configuration of the ultrafast gun assembly. The tip holder can be seen inside the cylindrical extractor electrode surrounding it. (D) Optical outline showing the focalization principle of the laser beam on the tip apex thanks to the combined action of the plan and the parabolic mirrors ($f' = 8$ mm). (E) Image of the FE Tip seen from the outside through the mirror optics inserted inside the gun and thanks to the CCD camera located in the optical head described in Fig. 2.18A.

designed to hold the flat and parabolic mirrors in the immediate vicinity of the tip as shown in Fig. 2.17B. The diameters of the flat and parabolic mirrors are, respectively, 8 and 6 mm. The design of the two mirrors and their holder, their compatibility with the UHV and high voltage requirements of the electron gun have been carefully optimized. In particular the mirror holder has been mirror polished with a final surface roughness of $R_a = 0.1\,\mu m$. Indeed, the mirror holder is finally inserted in the close vicinity of the V_1 electrode inner surface located only 2 mm below the mirror holder surface, as seen from the CAD design in Fig. 2.17C. The alignment of the mirror holder with respect to the tip is first realized outside the gun on a dedicated optical bench. The mirror holder is tightly held in position by a set of mirror polished screws and can then be moved with the tip during the final mechanical alignment of the electron beam with respect to the extractor hole. After this pre-alignment step the emitter, the mirror holder, and mirrors attached together with the tip holder are inserted in the gun assembly and sealed using the DN-CF40 flange. Air evacuation is then initiated followed by a 9 h baking and a careful high voltage conditioning of the accelerator to reach $V_0 = 200\,kV$. The laser beam is directed from the laser system to the tip using an optical setup mounted inside the so-called optical head (Caruso, 2019; Caruso et al., 2017). The small breadboard of this optical head is attached above the metallic plate located on the top of the metallic housing. The ray path inside this optical head before the beam enters the housing is outlined in Fig. 2.18A. The fine alignment of the laser beam on the tip apex is achieved by means of two piezo mirrors located in the optical head. A CCD camera and a white light are also located in the optical head allowing to see the tip when all the gun parts are sealed, baked and HV conditioned. A typical picture of the tip after installation and ready for operation is shown in Fig. 2.17E.

Thanks to a beam splitter located in the optical head, the laser beam can be injected and focused on the tip apex looking at the tip simultaneously using the white light. The beam of a red laser, used for the optical alignment, focused on the tip and captured by the CCD in the optical head is presented in Fig. 2.18A. The probe current I_p coming from the emission current generated by the interaction between the ultrafast beam and the W tip, can be acquired using a Faraday cup located in the condenser aperture plane right after the accelerator and before going through the illumination system of the microscope (Houdellier et al., 2018). Fig. 2.18B is a map of this probe current I_p obtained by scanning the position of the laser beam on the tip using piezo mirrors also located inside the optical head. It clearly shows that the

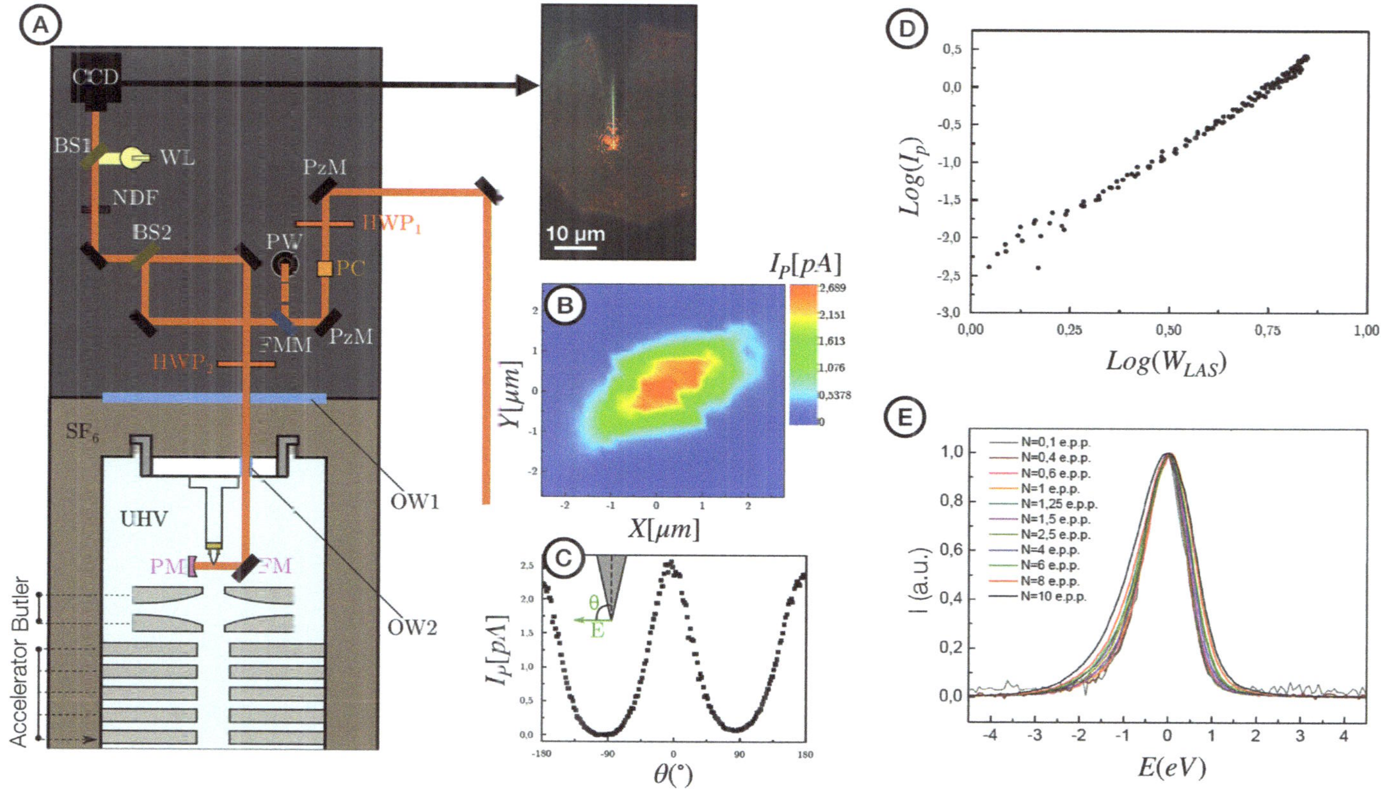

Fig. 2.18 (A) Optical head used to drive the laser beam into the Ultrafast CFEG: HWP, Half-Wave Plate; FMM, Flip Mount Mirror; PW, Power meter; PzM, Piezoelectric Mirror; BS, Beam Splitter; WL, White Light source; OW, Optical Window; FM, Flat Mirror; PM, Parabolic Mirror; PC, Polarizing Cube; NDF, Neutral Density Filter. (B) Map of the probe current measured by the Faraday cup located below the accelerator obtained by scanning the position of the laser beam on the tip apex using one PzM. (C) Probe current measured by the same Faraday cup for different polarizations of the femtosecond laser pulses focused on the *W* tip inside the electron gun. (D) Dependence of the probe current measured with the Faraday cup located below the accelerator, on the incident laser power. (E) Series of zero loss peaks (ZLP) acquired for different numbers of electrons per pulse (NEPP). *From Caruso, G.M., 2019. Development of a coherent ultrafast transmission electron microscope based on a laser-driven cold field emission source. Université Toulouse 3, Toulouse, PhD thesis.*

electron emission is restricted to a region having an extension smaller than 1 μm. The probe current reaches 2.8 pA in that case, with an incident laser power of 4.5 mW and a repetition rate of $f_{rep} = 1$ MHz. This corresponds to an average of ≈20 electrons per laser pulse inside the probe beam. Fig. 2.18C shows the polarization dependence of the probe current I_p. Maximum emission is obtained when the polarization of the laser electric field is parallel to the tip axis (Hommelhoff et al., 2006b). Fig. 2.18D shows the dependence of the probe current I_p on the incident average laser power W_{LAS}. The slope S extracted from the logarithmic curve $\log(I_p) = S \times \log(V_{LAS}) + B$ shows that electron emission is triggered from the apex of the nanotip by multiphotons field assisted photoemission, mainly two photons (Caruso et al., 2017; Ropers et al., 2007; Schenk et al., 2010). This makes sense regarding the photon energy $E_{photon} = 2.4$ eV given by the SHG ultrafast beam and the tungsten exit work function $\Phi = 4.5$ eV. To discriminate a possible heating contribution of the emitter from the laser beam, autocorrelation measurements have been also implemented and all the results are described in the Giuseppe–Mario Caruso thesis and in the literature (Caruso, 2019; Caruso et al., 2017). These measurements consist in sending a sequence of two laser pulses on the tip and measuring the probe current I_p as a function of the delay between the two pulses. They demonstrated that the electron pulse duration is limited by the laser pulse duration and that cumulative heating effects can be discarded. This conclusion is further supported by a theoretical estimation of the average temperature rise inside the tip apex which is of the order of 10 K.

The Zero Loss Peak (ZLP) has been also studied to determine the influence of the electrons pulse energy resolution ΔE as a function of the number of electrons per pulse (NEPP) (Caruso et al., 2019). NEEP is linked to the probe current I_p and is increased thanks to the incoming average laser power W_{LAS} following the power dependance reported in Fig. 2.18D. The resolution ΔE is measured using the full width at half maximum (FWHM) of the ZLP. Fig. 2.18E shows the evolution of the ZLP as a function of NEEP chosen between [0.1 and 10]. The energy resolution of the femtosecond electron pulses is in the eV range and increases with NEPP due to electron–electron interaction. However, even with NEEP<1, ΔE remains significantly larger than the one measured on a conventional CFEG (≈0.4 eV), with an ultimate measurement of $\Delta E = 1.08$ eV given by $NEPP = 0.1$. This is related to the ultrafast dynamics of the electrons gas inside the emitter just after the interaction with the laser pulse. Simulations reported during the Giuseppe–Mario Caruso PhD support this explanation and anticipate a reduction of this effect using single photon photoemission instead of the multi-photons process (Caruso, 2019; Caruso et al., 2019). It could be implemented by extracting the electrons pulse using UV pulses generated by fourth harmonic generation of the 1030 nm laser instead of the SHG.

Numerical simulations have been also performed to understand the influence of the mirror holder on the electron optical properties of the modified CFEG. Indeed, as seen in Fig. 2.17C, the mirror holder is located very close to the extracting electrode resulting in the modification of the electrostatic equipotentials distribution around the tip and the extractor hole. By computing the electrostatic field in the tip vicinity using FEM software COMSOL multi-physics we have first measured a 2.5 times reduction of the apex field when the mirror holder is inserted (COMSOL AB, n.d.). These simulations with and without the mirror holder have been carried out using same V_1 potential. In practice, this effect prevents the emission of continuous electrons beam using standard V_1 voltages ranging between [2 and 6] kV. As a result, after the mirror holder is inserted, a continuous electron beam begins only above $V_1 = 7\,\text{kV}$ (Caruso et al., 2017). The mirror holder being cylindrically symmetric, it does not affect the symmetry of the electric field at the tip apex. Using SIMION we have first cross-checked the previous results on the electrostatic field around the tip and then computed electron trajectories with and without the mirror holder for the total Butler lens including the focusing electrode V_2 (Dahl, 2000). The two results are reported in Fig. 2.19B. Our calculations have consistently shown that the mirror holder shrinks the electron trajectories toward the source axis. This confinement can be clearly seen by comparing the left and right sides of Fig. 2.19B. Paraxial properties as well as primary aberrations have been also computed using EOD again for the total Butler lens (Lencová and Zlámal, 2008). The two simulations with and without mirror holder have been also inserted in Fig. 2.19A. Looking at the position of the marginal ray virtual cross-over of the Butler lens, we observe an increase of the distance between this cross-over and the tip due to the mirror holder. This change in the paraxial behavior of the Butler lens is equivalent to the trajectories confinement computed with SIMION. Aperture as well as chromatic aberrations have been also determined. Carrying out these simulations using the same voltages conditions ($V_1 = 7\,\text{kV}$ and $V_2 = 35\,\text{kV}$) for the two Butler lens models show no strong detrimental change on the aberrations coefficients due to the mirror holder. In conclusion, the mirror holder only affects the paraxial properties of the Butler lens without any additional contribution to the primary aperture and chromatic aberrations. This small change in the paraxial strength of the gun electrostatic lens can be easily compensated using

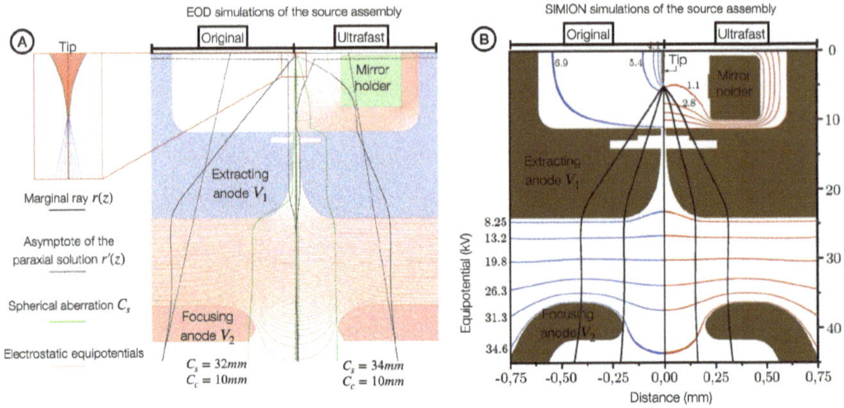

Fig. 2.19 Simulations of the optical properties of the modified ultrafast gun assembly containing the tip, the extracting electrode V_1 and the Butler lens V_2/V_1. (A) Paraxial marginal trajectory calculated using EOD: comparison between the standard gun assembly (*left*) and the modified one (*right*). The position of the paraxial virtual image cross over as well as the evolution of the aperture aberration C_s are also reported. (B) Simulation of two axial rays using SIMION: comparison between the standard gun assembly (*left*) and the modified one (*right*). Such as for EOD, the electrostatic equipotentials are also reported with their corresponding value in kV (*left vertical scale*). *From Caruso, G.M., Houdellier, E., Abeilhou, P., Arbouet, A., 2017. Development of an ultrafast electron source based on a cold-field emission gun for ultrafast coherent TEM. Appl. Phys. Lett. 111 (2), 023101.*

the ratio $R = V_2/V_1$. The evolution of ultrafast electron probe depending on the ratio is reported in Fig. 2.20A. Below $R = 6$ the Butler lens generates a virtual crossover located before the tip, as previously described in Fig. 2.5 for standard HF2000 CFEG. With the mirror holder inserted, the shift between virtual and real cross over conditions is observed around $R = 6$ instead of $R = 7$ in conventional CFEG. This is a direct consequence of the Butler lens paraxial changes computed using EOD (Caruso, 2019).

Looking at Fig. 2.20A, we could observe the evolution of the ultrafast source demagnification in virtual and real cross over conditions. Indeed, the probe size increases (*i.e.* the lens demagnification decreases) when the Butler strength get closer to $R = 6$, and decreases again (*i.e.* the lens demagnification increases) from $R = 6.9$ to $R = 8$. We have also determined the time-averaged probe brightness defined by:

$$B_{av} = \frac{f_{rep} N_e e}{(\pi r^2)(\pi \alpha^2)} \tag{2.10}$$

in which f_{rep} is the laser repetition rate, N_e is the NEPP, e is the elementary charge, r is the radius of the electron probe and α is the beam convergence semi-angle. This probe brightness will be equivalent to the real source

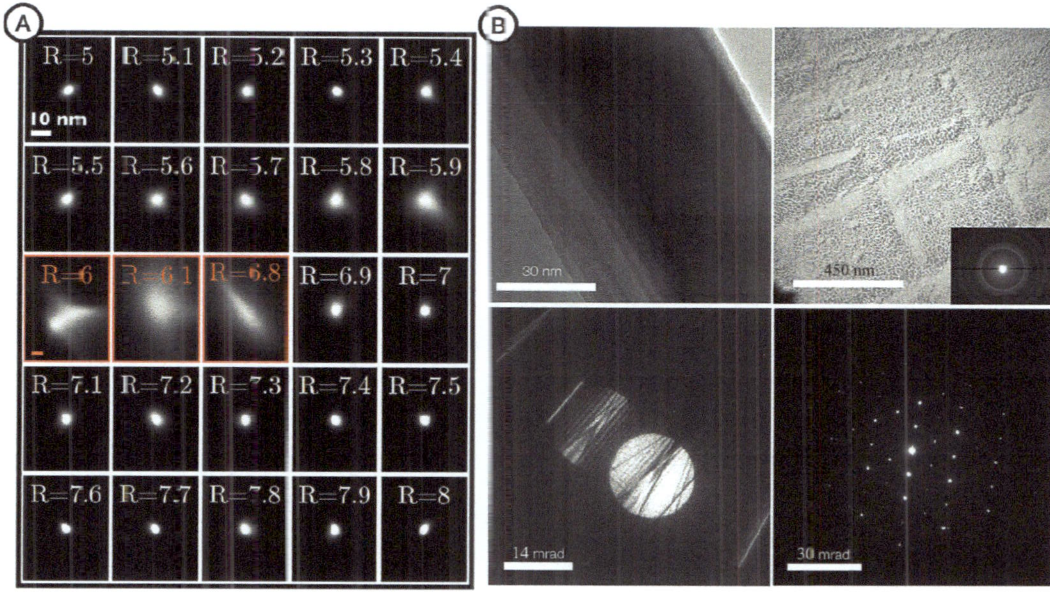

C

Probe current in standard FE-TEM	Probe current in ufast FE-TEM
100pA - 1nA	0.1fA - 100fA
TEM Technique	**Exposure time (s)**
Imaging	100
HREM	150
SAED	150
CBED	100
EELS	Low loss :1 core loss 60

Fig. 2.20 (A) Ultrafast electron spots acquired for different ratios R showing the transition between virtual to real gun crossover position performed by the Butler lens. This transition is located around $R \approx 6$ (conditions: repetition rate $f_{rep} = 2\,\text{MHz}$, average number of electrons per pulse $N_e = 12.5$, integration time 20 s). (B) *Top-left*: Ultrafast HREM image of a crocidolite crystal ($f_{rep} = 2\,\text{MHz}$, $N_e = 11$, 150 s, Binning 4). *Top-right*: Ultrafast conventional BF image of a gold crossgrating ($f_{rep} = 2\,\text{MHz}$, $N_e = 11$, 100 s, Binning 1). *Bottom-right*: ultrafast SAED pattern taken along the [110] direction of a TiAl γ-phase crystal ($f_{rep} = 1\,\text{MHz}$, $N_e = 12.5$, 150 s, Binning 2). *Bottom-left*: ultrafast CBED pattern acquired in (220) two beams condition using a Si crystal ($f_{rep} = 1\,\text{MHz}$, $N_e = 12.5$, 100 s, Binning 1). (C) Table summarizing the difference of probe currents available inside a conventional FETEM and the one modified to generate ultrafast electrons pulses. The typical acquisition times required for TEM techniques using the ultrafast CFEG are also reported in C (see Houdellier et al., 2018).

brightness only in paraxial or aplanatic condition for the illumination optic (Fransen et al., 1999a). Experimentally we tried to be the closest possible to the paraxial conditions using a small STEM aperture of 30 μm. The time-averaged brightness in ultrafast emission mode has been determined both in real ($B_{av} \approx 2.5 \times 10^3$ A cm^{-2} Sr^{-1}) and virtual ($B_{av} \approx 5.8 \times 10^3$ A cm^{-2} Sr^{-1}) cross-over conditions of the gun. These values are the highest measured brightness among all UTEM instruments reported in the literature until now (Feist et al., 2017; Houdellier et al., 2018; Zewail and Thomas, 2010). Considering $f_{rep} = 1$ MHz of laser repetition rate, electron emission occurs during $\approx$ 350 fs, *i.e.* the laser pulse duration, every 1 μs. Therefore, assuming a constant value of the instantaneous current, the measured value would correspond in continuous emission to a brightness in the 10^9 A cm^{-2} Sr^{-1} range, *i.e.* similar to state-of-the-art continuous CFEG. Finally an angular current density of 2 nA Sr^{-1} has been determined. Combining these measurements together we have estimate a virtual source size in the 5–10 nm range, which is the smallest achieved to date in UTEMs and equivalent to standard CFEG virtual source size (Hawkes, 2009). Therefore, our ultrafast CFEG allowed to generate bright ultrafast electrons pulse at the expense of a strong decrease in the probe current. Unfortunately this current loss will have a significant impact on the capabilities of our coherent UTEM to implement sensitive methods such as electron holography. The main motivation for my research project, which will be outlined in the last chapter, will be to address this issue.

The potential of the new TEM equipped with the ultrafast CFEG, illustrated in Fig. 2.20B, has been investigated for conventional and high-resolution imaging as well as electron diffraction in parallel (SAED) and convergent beam configuration (CBED) (Houdellier et al., 2018). The UTEM was operated using an acceleration voltage of 150 kV. A bright field image of a gold cross grating, used as a calibration sample, has been acquired using the ultrafast emission and is reported in the top right part of Fig. 2.20B. Additionally the high-resolution image of a crocidolite crystal is shown in the top left part of Fig. 2.20B where the 0.9 nm lattice planes could be separated. A SAED pattern taken along the [110] direction of a TiAl γ-phase crystal and the ultrafast CBED pattern obtained in (220) two beams orientation in a silicon crystal are also reported in the two micrographs located at the bottom of Fig. 2.20B. These images and diffraction patterns obtained with ultrashort electron pulses (duration smaller than 400 fs at the sample) are of comparable quality as the ones obtained with the HF2000 in conventional mode. Moreover, the excitation of all lenses was the same in ultrafast and conventional mode. In ultrafast mode, we could then measure a lattice image resolution of 0.9 nm at 150 kV, thanks to the crocidolite crystal. However, the atomic structure of gold particles in the cross grating sample could not be resolved in ultrafast mode while it can be easily resolved using standard HF2000. This limitation is a direct consequence of the strong reduction in available probe current arising from the ultrafast nature of the emission.

Indeed, in conventional continuous CFEG the emission current is usually set around 10 μA, which will generate a probe current of approximately 1 nA (measured with illumination set in STEM mode using a 30 μm STEM aperture). This four order of magnitude difference between emission and probe current is due to the aberrations of the illumination system, and mainly to the one coming from the Butler lens system introduced in Section 2.3.2 and described in Fig. 2.3 (Hawkes, 2009). Usually, the influence of these aberrations are minimized using small condenser (or STEM) apertures preserving the original source brightness in the probe at the expense of this 10^4 loss in the final probe current. In continuous emission this difference is not really a problem as 10 μA of emission current is available in the source and 1 nA of probe current is largely sufficient for almost all analysis.

However, in ultrafast mode the same electron optical condition ($V_1 = 4$ kV and $R = 5$) will yield an emission current around 2.5 pA for an incident average laser power of 6 mW and a repetition rate of $f_{rep} = 1$ MHz. This is equivalent to approximately $N_e = 15$ electrons per pulse. Using the same illumination setup than before, the corresponding 10^4 loss between emission and probe will then generate a probe current around 0.2 fA. Traditionally, we choose to use a bigger STEM aperture to reach ≈50 fA of probe current even if the brightness will be slightly deteriorated. This will correspond to an electron dose on the sample of approximatively 5–20 electrons per Angström square equivalent to the dose traditionally set to observe extremely fragile sample in conventional TEM imaging *i.e.* the low dose mode of TEM (Yakovlev and Libera, 2008). These ultrafast probe current values are limited by the tip damage threshold and the laser repetition rate. Therefore, the exposure time has to be increased and the image resolution remains mainly limited by the microscope stability during the exposure time and the detector properties. This is why gold atomic structure could not be resolved using the ultrafast emission as HREM imaging requires 150 s of exposure time. Fig. 2.20C reports a list of typical exposure times used for the acquisition of good TEM micrographs using various techniques implemented thanks to the ultrafast emission. Improving this emission/probe current loss thanks to innovative source optical setup will be the direction taken by my research project which will be described in the last chapter.

2.5.3 Moving toward coherent UTEM

I will now briefly describe two applications of this new coherent UTEM showing the potentialities of our developments even using low dose like probe current. The first one called electron energy gain spectroscopy (EEGS) is a pump-probe experiments set to determine the temporal coincidence between the laser pump and the electron probe on the sample plane (Caruso et al., 2019). EEGS is central for future UTEM experiments since it must be always carried out before any other experiments to determine the starting time $\Delta t = 0$ given by a specific position of the delay line. Sections 2.5.3.1 and 2.5.3.2 will look at how our ultrafast CFEG may be used to do off-axis electron holography experiments despite the low probe current, paving the way for future ultrafast holography studies with the pump-probe configuration (Houdellier et al., 2019).

2.5.3.1 Implementation of electron energy gain spectroscopy using pump-probe setup

Besides to the development of the ultrafast CFEG presented in the previous parts, the implementation of the pump-probe setup described in Fig. 2.13B requires the modification of the objective lens allowing to inject and focus the ultrafast laser beam on the sample. Using a free port located in the objective lens yoke, previously dedicated to the baking system of the OL pole piece, we introduced a mechanical setup originally developed in LPS-Orsay for cathodoluminescence experiments in a dedicated STEM VG microscope (Kociak and Zagonel, 2017; Kociak et al., 2018). This system has been adapted for the HF2000 in close collaboration with the colleagues from LPS (Houdellier et al., 2018). It is composed of a parabolic mirror placed above the specimen holder inside the 4.5 mm pole piece gap of the HF2000. The mirror has a focal distance $f = 1$ mm and a 300 μm hole allowing the electron beam to pass through. It can be moved from outside by a XYZ translation stage equipped with three micrometrical screws. Carrying out the injection of the laser beam from the outside to the inner area of the objective lens and allowing it's focalization using the parabolic mirror enabled us to excite the sample in the same area probed by the electrons after an adjustable delay Δt (see Fig. 2.13B). A computer controlled mechanical delay line is installed in the laser pump line located on the optical table close to the microscope. By moving this delay line, the length of the optical pump line can be changed very precisely compare to the probe one allowing the delay Δt to be adjusted.

Electron energy gain spectroscopy (EEGS) is a UTEM technique that characterizes the optical or vibrational excitations of nanostructured systems from the detection of electrons which have gained energy during their interaction with the sample. The coupling between electrons and photons in free space is forbidden due to the energy-momentum mismatch. However, electrons and photons can couple in free space by mean of a third system or a particle (de García, 2010; de García et al., 2010). For instance, an electromagnetic wave incident on a nanostructure creates an evanescent electric field that can provide the necessary extra momentum to couple electrons and photons as outline in Fig. 2.21A. In particular, the field component parallel to the electron motion yields efficient interactions between electrons and light. Implementing this setup experimentally has been realized using a regular tungsten tip located in the sample plane of the HF2000. This tip will then act as a nanostructure, coupling electrons and photons. The electron beam is focused closed to the edge of the tip where the optical near field component will be stronger (see Fig. 2.21A). Fig. 2.21B shows a typical EEGS/EELS spectrum acquired when the delay $\Delta t = 0$ (bottom one) and a zero loss peak (ZLP) given by the electron probe with a delay larger than the pulses time resolution $\Delta t >> \Delta t_{pulse}$ *i.e.* without any interaction between the field and the electrons arriving too late (top spectrum). The pump wavelength is 515 nm and the pulse average power measured on the external optical bench was 595 mW. Bidimensional spectra are displayed in Fig. 2.21B where the horizontal direction stems for the energy scale and the vertical for the angular one defined by the spectrometer entrance aperture (SEA) set at 1 mm. The microscope has been also adjusted in diffraction mode. Angular averaged spectrum along the vertical direction are also reported below the bidimensional ones.

From the comparison between the two spectrum, we can notice that when $\Delta t = 0$ the ZLP is depleted and sidebands appear (Caruso et al., 2019; Feist et al., 2017; Yurtsever and Zewail, 2012). The latter are populated by electrons that absorbed or emitted photons during the interaction with the evanescent electric field (de García et al., 2010). The energy spacing between the peaks is equal to the incident photon energy $E_{photon} = \hbar\omega$ (2.4 eV in our case). EEGS could then be a powerful technique allowing to obtain informations on the local electric field in proximity of a nano-sized system illuminated by a laser pulse (Yurtsever and Zewail, 2012). However, we will now be more interested in its ability to determine the temporal features of the electrons pulse.

A sequential acquisition of 52 EEG spectra with a step of 50 fs in the delay Δt has been realized and interspersed with the same number of ZLP acquisitions to correct a posteriori a drift of the spectrometer. Each spectrum has

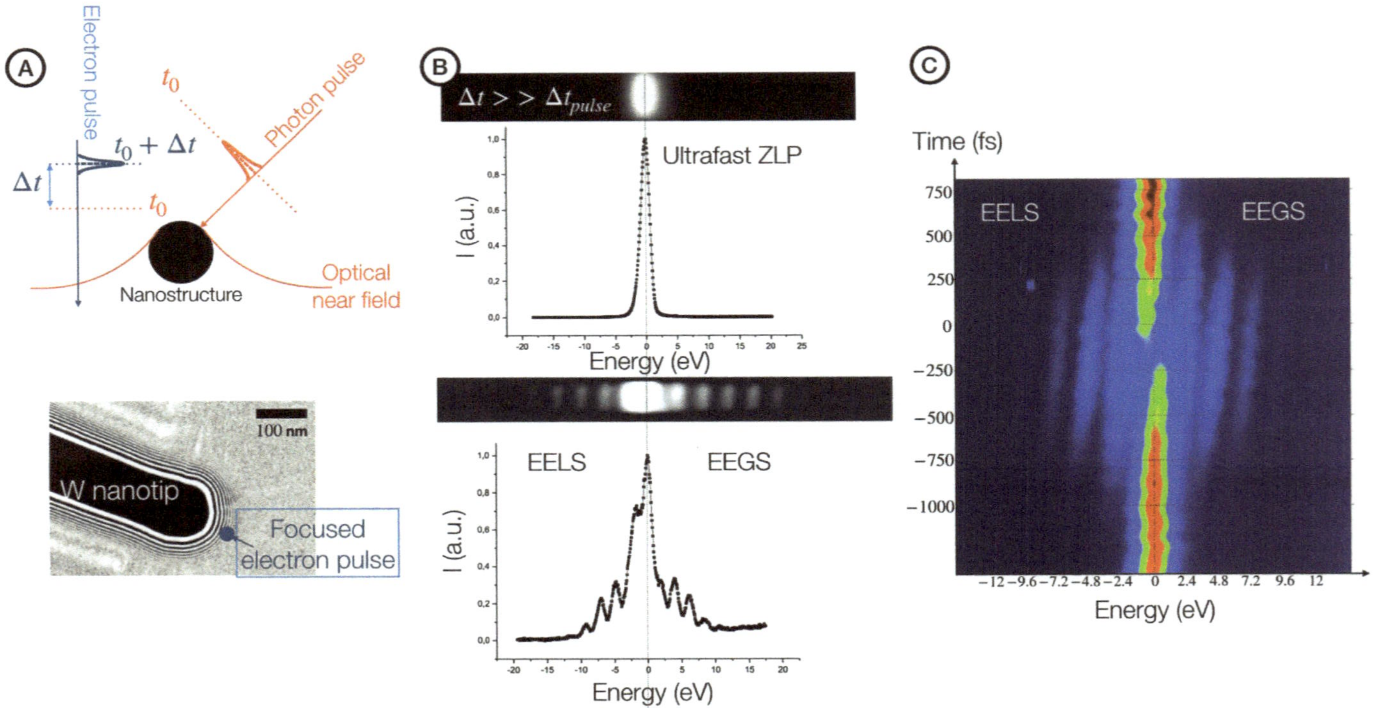

Fig. 2.21 (A) Schematic representation of the electron–photon interaction obtained thanks to the presence of a nano-system. An evanescent electric field will be generated from the interaction between the photon pulse and this nano-sized object. The evanescent field will then interact with the ultrashort electrons pulse after an adjustable delay Δt. In our experiments, the selected nano-system was the apex of a *W* tip. (B) Electron energy loss (EELS) and gain (EEGS) spectrum acquired from the ultrashort electrons pulse after it's interaction with the evanescent field located at the edge of the *W* apex. *Top*: acquisition performed using a too large delay *i.e.* the electrons pulse and the evanescent field generated by the photons pulse are not superimposed in time. *Bottom*: acquisition performed using a delay close to $\Delta t = 0$ maximizing the interaction between the electrons pulse and the evanescent field in time. (C) Temporal scanning of the EELS/EEGS spectrum obtained varying the delay Δt between the photon pump and the electron probe. *From Caruso, G.M., 2019. Development of a coherent ultrafast transmission electron microscope based on a laser-driven cold field emission source. Université Toulouse 3, Toulouse, PhD thesis.*

been acquired using a $t = 10\,\mathrm{s}$ of exposure time, $SEA = 1\,\mathrm{mm}$ and the energy dispersion 0.05 eV/channel. Fig. 2.21C reports all the spectra as a function of the time delay Δt. The variation of the electrons population as a function of the time delay is clearly visible from Fig. 2.21C. When the electron and laser pulses are synchronized, the ZLP depletes and the electrons populate the sidebands. The electron pulse time resolution Δt_{pulse} could be evaluated by comparing the range of time delay Δt during which sidebands are observed in Fig. 2.21C, with the range determined theoretically by simulating the time autocorrelation between the photon pump pulse (known) and the electron pulse one (unknown). Of course Δt_{pulse} will also depend on the NEPP. A rigorous evaluation has been carried out by Giuseppe–Mario Caruso during his PhD thesis (Caruso et al., 2019). All theoretical details as well as experimental curves acquired as a function of the NEPP can be found inside his manuscript (Caruso, 2019). For instance, considering a photon pulse of 350 fs, he found a FWHM of $\Delta t_{pulse} = 400\,\mathrm{fs}$ for the case $N_e = 0.59$ with a precision of 6%. The pulse duration increases to $\Delta t_{pulse} = 524\,\mathrm{fs}$ for the case $N_e = 4$ reported in Fig. 2.21C.

Additionally, the presence of a chirp effect in the electron pulse can be identified by the tilt of the sidebands. Indeed, this effect, which depends on the energy spread and the NEEP, is usually understood as the time dependence of the electron energy *i.e.* the electrons having more energy will arrive on the sample faster than the one with less energy (Hui and O'Sullivan, 2009). The chirp dependance to the NEPP has been also studied in detail by Giuseppe during his PhD. He found a chirp increasing almost linearly from $1100\,\mathrm{fs\,eV^{-1}}$ for $N_e = 0.59$ to $825\,\mathrm{fs\,eV^{-1}}$ for the case $N_e = 4$ shown in Fig. 2.21C (Caruso, 2019).

2.5.3.2 Off-axis electron holography using ultrafast emission

As already stated in Section 2.5.2, there is a difference of approximately three to four orders of magnitude in the emission current between continuous and laser-driven mode, under the same electron optical conditions. In the latter the current is also reduced due to the effect of the laser repetition rate f_{rep} and the limited NEPP allowed by the tip and by the time properties required for the pulse. Looking at the good average brightness of the ultrafast CFEG it should be possible to implement off-axis holography in UTEM but the small probe current will strongly influence the phase detection limit $\delta\varphi$ defined in Eq. (1.42). The number of electrons per pixel N in Eq. (1.42) can be determined by the relation:

$$N = N_{ppx}.f_{rep}.t_{exp} \tag{2.11}$$

where f_{rep} is the laser repetition rate, t_{exp} the exposure time and $N_{ppx} = \frac{N_e}{N_{px}}$ the number of electrons per pulse per pixel. N_e indicates the NEPP and N_{px} the total number of pixel in the measurement area. Substituting Eq. (2.11) in Eq. (1.42), and considering Eq. (1.43), we obtain:

$$\delta\varphi = \frac{(snr)_\varphi}{\gamma_{total} \times C_{inst} \times MTF} \sqrt{\frac{2.N_{px}}{DQE \times N_e.f_{rep}.t_{exp}}} \tag{2.12}$$

assuming the contribution of inelastic scatterings to be negligible. The number of electrons emitted per laser pulse is:

$$N_e = \frac{I_P}{e.f_{rep}} \tag{2.13}$$

I_P being the probe current and e the elementary charge. Knowing that our emission process is a multi-photon process involving n photons *i.e.* $I_p \propto W^n_{LAS}$, we can easily explicit the relation between $\delta\varphi$ and the average laser power, W_{LAS}:

$$\delta\varphi \propto \frac{(snr)_\varphi}{\gamma_{total} \times C_{inst} \times MTF} \sqrt{\frac{2.e.N_{px}}{DQE \times \ W^n_{LAS}.t_{exp}}} \tag{2.14}$$

As in continuous mode, the spatial coherence $1D6FE_{total}$ is maximized with the highest brightness source and using an appropriate illumination setup, while the MTF depends on the detector and the parameters set during the experiment (*i.e.* camera binning, hologram magnification, spatial frequency, biprism voltage). In Eqs. (2.12) and (2.14) there are new parameters related to the optical excitation. By changing the laser power W_{LAS}, it is possible to adjust the number of electrons per pulse N_e. However, the latter is limited for two reasons: the cathode damage threshold and the probe properties rapidly deteriorate increasing NEPP, as described in Section 2.5.3.1. The laser repetition rate f_{rep} in Eq. (2.12) cannot be freely changed to improve the phase detection limit because it is limited by the dynamics of the specimen to be investigated.

The exposure time is another free parameter that can be easily adjusted. As in the case of holography performed with a continuous beam, an increase of t_{exp} can improve $\varphi\delta$ by increasing N_e at the expense of higher instabilities contribution, *i.e.* C_{inst} in Eqs. (2.12) and (2.14).

Fig. 2.22A shows a series of off-axis holograms acquired in the vacuum with increasing exposure time t_{exp}. We can notice that the case $t_{exp} = 100\,s$

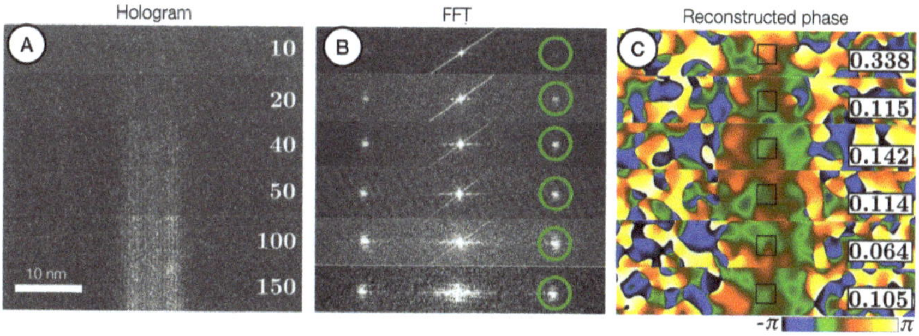

Fig. 2.22 Influence of the exposure time on the hologram. (A) Electron holograms acquired in vacuum. Exposure times in seconds are displayed in *white*. (B) Power spectra of the holograms. *Green circles* indicate the mask position in the Fourier space used for the phase reconstruction. The achieved spatial resolution is 2 nm. (C) Reconstructed phases. Standard deviations calculated inside the *black square* are displayed. Experimental conditions: biprism voltage 18.8 V, W_{LAS} = 8 mW, f_{rep} = 2 MHz, $N_e \approx 9$, binning 1, magnification 200 kX. *From Houdellier, F., Caruso, G.M., Weber, S., Hÿtch, M.J., Gatel, C., Arbouet, A., 2019. Optimization of off-axis electron holography performed with femtosecond electron pulses. Ultramicroscopy 202, 26–32.*

reports the lowest standard deviation σ_φ. Using longer exposure times standard deviations start to deteriorate due to the influence of instabilities (Houdellier et al., 2019).

100 s of exposure time is still too high for conventional holography application requiring higher field of view. Therefore, we have implemented the acquisition of hologram stacks. Image stacks consist in a sequential acquisition of several holograms. Using a stack of a high number of holograms with short exposure times allows to correct the effect of instabilities. However, due to the latter, simply summing all the hologram fringes would not improve the contrast. The resulting hologram would corresponds to a standard one acquired with a total exposure time equal to $t_{exp} \times N_{holo}$, where t_{exp} is the acquisition time of each individual hologram and N_{holo} the number of holograms in the stack. In order to properly exploit hologram stacks, a realignment process is required. Data processing has been performed using home-made scripts developed in CEMES by Christophe Gatel (Boureau et al., 2018; Gatel et al., 2018). These methods were originally implemented during the Julien Dupuy PhD thesis to align automatically the holograms during their acquisition in the I2TEM (Dupuy, 2021). The full description of the realignment process can be found in the Giuseppe's manuscript too (Caruso, 2019). Fig. 2.23A shows the effect of the individual hologram exposure time and the threshold of instability contribution in the hologram contrast.

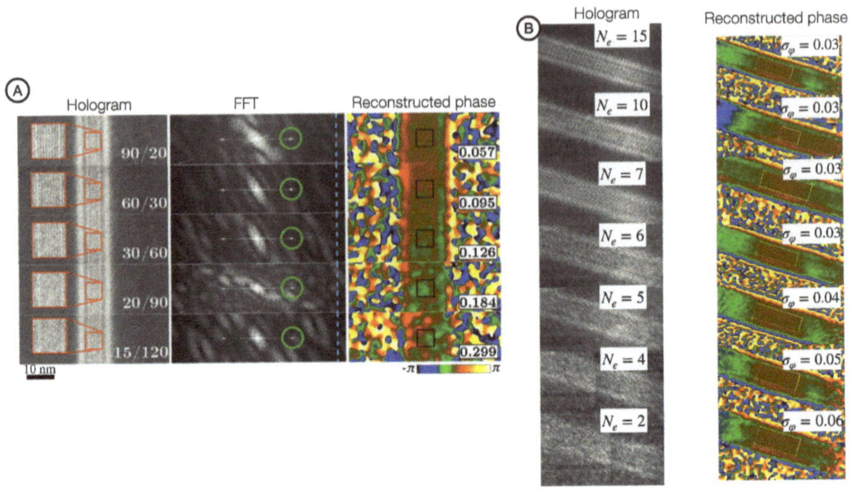

Fig. 2.23 (A) The effect of the individual hologram exposure time in summed corrected hologram stacks. The total exposure time remains constant. *Left*: off-axis holograms acquired in the vacuum (the white numbers indicate the number of images in the stack and the individual acquisition time). *Center*: Power spectra of the holograms. The *green circles* represent the 2.5 nm resolution mask in Fourier space used for the phase reconstruction. The *blue dashed line* shows the position of the Nyquist frequency. *Right*: reconstructed phases. Standard deviations calculated inside the *black square* are displayed. Experimental conditions: $W_{LAS} = 8\,\text{mW}$, $f_{rep} = 2\,\text{MHz}$, $N_e \approx 6$, binning 2. Magnification: 400 kX. Biprism voltage: 30 V. (B) Influence of the NEPP in off-axis electron holography. All holograms are extracted after a careful realignment from the stack of 90 single holograms acquired with 20 s of exposure time each (same conditions as before). The associated reconstructed phase is also reported on the right with the evolution of the phase standard deviation. *From Houdellier, F., Caruso, G.M., Weber, S., Hÿtch, M.J., Gatel, C., Arbouet, A., 2019. Optimization of off-axis electron holography performed with femtosecond electron pulses. Ultramicroscopy 202, 26–32.*

The total exposure time remains constant at 1800 s. We can clearly see that by increasing the individual acquisition time, the contribution of instabilities becomes more important. Hence, stacks composed of a high number of individual holograms acquired with shorter exposure times should improve even more the contrast. However for exposure times shorter than 20 s, the hologram contrast is not sufficient for the correction algorithm to be used properly. Fig. 2.23B also shows the influence of the NEPP. Obviously, the use of higher NEPP decrease the standard deviation at the expense of a loss in the time properties of the electron pulse.

Carrying out off-axis electron holography experiments using ultrafast emission is then possible. Standard deviations and phase detection limits can be improved implementing stack acquisition of 20 s holograms and a careful realignment procedure between each individual hologram which are finally summed up. An optimum standard deviation of $\sigma_{\varphi} = 0.03$ rad has been obtained for an hologram field of view of $W = 10$ nm and a NEPP of $N_e = 15$. The implementation of these methods to measure the mean inner potential of a MgO nano-cube test sample can be found in the Giuseppe–Mario Caruso PhD's manuscript (Caruso, 2019). The corresponding phase detection limit determined under ideal conditions ($(snr)_{\varphi} = 3$, $C_{inst} = 1$) was $\delta\varphi = 0.1$ rad which is a reasonable number regarding future applications (Houdellier et al., 2019).

The remaining limitation to study the ultrafast dynamic of a sample performing time resolved pump-probe off-axis electron holography will be the too small hologram field of view which cannot be easily improved in single biprism condition without Lorentz mode. This technical limitation cannot be simply overcome inside the HF2000 prototype. A second important limitation is coming from the sample drift arising during the stack acquisition. The previous realignment procedure only concerns the hologram fringes, but the sample has to be realigned between each individual hologram as well. This will require to control the sample stage during the stack acquisition. Fortunately, these kind of corrections have been already implemented for conventional holography experiments using the I2TEM during the PhD work of Julien Dupuy and all the details can be found in his manuscript (Dupuy, 2021; Gatel et al., 2018). As described in Section 1.3.2, the I2TEM was indeed designed to optimize holography experiments and such a dedicated column could then solve all the remaining limitations for ultrafast holography at once. Therefore, implementing ultrafast characterization of materials using time-resolved off-axis holography will require to transfer this ultrafast CFEG source on the I2TEM or an equivalent microscope. This second option will be implemented on a dedicated HF3300 column within a joint laboratory created between HHT and the CNRS called HC-IUMi for Hitachi-CNRS Infrastructure for Ultrafast Microscopy (Houdellier and Arbouet, 2018). This will be briefly described in the following part.

2.6. Discussion and introduction to the research project

All along my research work, I was interested in the development of new coherent electron microscopy methods. In the previous sections, I choose to separate my contribution to this field in two distinct parts: the methodological approach and the instrumental one. Dealing with coherent techniques such as highly focused CBED or off-axis electron holography requires specific optical configurations of the microscope column. The development of the I2TEM together with HHT enabled us to implement optimized off-axis electron holography experiments and also to develop new kind of interferometric setups such as SCBED or TRW-DFEH. Optimizing these complex configurations have required the development of paraxial simulation codes for the calculation of fundamental rays as well as ray tracing simulations of the full microscope using SIMION. All these developments are part of the methodological approach which aims to reach the full potential of a given TEM optic such as the I2TEM.

By modifying key components of the CFEG I also tried to improve the capabilities of coherent techniques from the root. For instance, decreasing the source size using innovative cathode materials I tried to improve the phase detection limit in holography or the probe brightness in STEM; then enabling the electron emission to be triggered by a femtosecond laser I allowed the generation of ultrashort bright electrons pulse enabling the coherent methods to be implemented in a time-resolved fashion.

These two research directions should continue to be explored. These new developments, on the other hand, will correspond to the logical next steps in my prior research and cannot be deemed fresh directions for the research project described in the third chapter. I could then briefly described these following developments as follow:

1. Methodological developments: we will continue the work initiated by Yudai Kubo from HHT and strongly upgraded by Julien Dupuy during his PhD. We will improve the I2TEM real tracing simulation code computed within SIMION and also the visualization of the calculated trajectories. In particular, I would like to implement the determination of all lenses nonlinear properties thanks to the real tracing simulation and also to work on a method allowing optical configuration optimization using paraxial and real tracing codes. Carrying out a configuration optimization will need initial requirements from the users (for instance: hologram field of view W, resolution s, ...) and, thanks to the paraxial solutions, the code should be able to find a first rough solution of the ideal microscope configuration fulfilling these requirements. Then, knowing all aberrations properties, the solution will have to be refined to determine the optimum excitations which will be inserted in the microscope lenses.
2. Instrumental developments: we will continue working on the transfer of the ultrafast CFEG on the HF3300 microscope within the HC-IUMi joint laboratory. Such as the I2TEM, the IUMi microscope is also

equipped with a double stage objective lens allowing the implementation of Lorentz imaging configurations described in Section 1.3.2, except the TL_{11} Lorentz mode as the column will not be equipped with a B COR. The conversion of this microscope in a new UTEM will require the production of a new 300 keV metallic housing, two laser injectors *i.e.* one for the Lorentz sample stage and a second one in the normal stage position. All these modifications will be strictly identical to the one performed on the HF2000 previously described, but adapted to a 300 keV column. The final IUMi UTEM will enable the implementation of time-resolved electron holography performed using single focus or double focus Lorentz imaging modes with the opportunity to use the double biprism configuration. Such as the I2TEM optic, which has enhanced conventional holography capabilities compare to SACTEM, the IUMi UTEM should open new perspectives in time-resolved electron holography originally tested on the HF2000 prototype. Nevertheless, the IUMI microscope will remain limited by the small probe current, such as the HF2000 prototype. Indeed, since these limitations originate mainly from the Butler lens properties, they will be exactly identical in IUMi equipped with the same Butler lens as the one used in the HF2000 prototype. This is why these works should be viewed as a technological transfer to a more efficient optical equipment rather than a brand new instrumental development that will overcome this limitation.

As a result, in the third chapter I would like to focus the discussion on a new instrumental direction I wish to follow. The concept behind this next step will be to introduce innovative electrons optics designs inside the CFEG to see beyond the traditional electrostatic Butler lens. Indeed, by improving the source optic we could tackle many limitations at once, for instance the demagnification limitation encountered using CcNT emitter or the limited probe current given by the ultrafast CFEG technology. These two problems are endemic to the poor optical properties of the Butler type electrostatic lenses. Moreover, improving CFEG optical properties should also lead to valuable consequences for other applications requiring an optimization of the probe brightness such as new monochromated CFEG. In the following chapter I will outline the concepts I would like to develop and test on the HF2000 bench.

CHAPTER THREE

Every electron counts: Toward the development of aberration-optimized and aberration-corrected electron sources[a]

Contents

Quand on travaille pour plaire aux autres on peut ne pas réussir, mais les choses qu'on a faites pour se contenter soi-même ont toujours une chance d'intéresser quelqu'un.

Marcel Proust

[a] *Coherent electron microscopy: contributions and project.* Florent Houdellier, CEMES-CNRS.

Advances in Imaging and Electron Physics, Volume 227
ISSN 1076-5670
https://doi.org/10.1016/B978-0-443-19324-8.00003-1

3.1. Introduction

The development of the first coherent UTEM opens up a wide range of opportunities such as the implementation of coherent techniques, like highly localized CBED, or off-axis electron holography, in a time-resolved fashion. In the previous chapter, we have introduced the first results obtained using holography and discussed future implementations of time-resolved holography experiments using the IUMi microscope. CBED geometry seems also very promising to be implemented using our UTEM. Indeed, by recording the time evolution of the HOLZ line rocking curves in a CBED pattern, we should be able to look into new structural informations in the ultrafast physics of nano-materials. For instance, by focusing a laser on a thin crystalline lamella, we could start a damped oscillation mechanism which can be precisely recorded in time using the pump-probe setup. The oscillation of a thin lamella will be observed through a time-dependent evolution of its curvature. We have previously demonstrated that a thin and curved crystalline lamella will exhibits an inhomogeneous strain field (Houdellier et al., 2008c). Yet, as described in Section 1.2.1, we know that such an inhomogeneous strain field along the propagation direction of the electron beam will be detected though the broadening of the HOLZ line rocking curves in a CBED pattern. Therefore, carrying out ultrafast CBED pump-probe experiments will allowed us to follow the time-dependent evolution of each HOLZ line rocking curve located inside the transmitted CBED disk. These rocking curves could then be quantitatively analyzed using the time-dependent dynamical theory approach developed during my PhD, and reported in Section 1.2.1.3. Combined with FEM simulations of the crystal strain state, it will be possible to retrieve the full four-dimensional strain tensor of the nano-structure, i.e., the time-dependent three-dimensional strain tensor of the crystalline lamella. Applying these methods to a perfect crystal with a specific shape, such as well-known MgO cubes, could be used to determine elementary vibrational properties of these nano-structures, i.e., the phonons and their dispersion properties. It would be, therefore, interesting to compare these results with recent phonons mapping obtained using monochromated STEM-EELS experiments (Egoavil et al., 2014). Indeed, from the physics point of view, studying structural dynamics of nano-materials using *very low energy* loss EELS, or *very fast* time resolved diffraction should be strictly equivalent. However, thanks to the three-dimensional nature of the CBED pattern, the diffraction configuration gives us much more information than a standard EELS experiment. We can see how a seemingly innocent thought experiment could already yield new interesting informations helping us to better understand the fundamental dynamic of nano-materials. This shows how important it is to continue investigations following that fruitful research direction.

However, I am also strongly convinced that the lack of probe current, intrinsic to the poor optical properties of the gun focusing optics of our first ultrafast CFEG design, must be addressed if we wish to expand the spectrum of applications of coherent UTEM while simultaneously improving the coherent source utilizing new cathodes. This is the reason why I have decided to follow this new direction for my research project: *designing and implementing original electron optical setups to improve the probe brightness of FE-TEM instruments*. In order to address this project with the appropriate tools, our previous work has shown that it is necessary to first spend time on the development of an optimal strategy for the simulation of a charged particle optical system. Ideally, this should also include a program for optimizing the mechanical parameters with respect to the targeted optical properties. Before going deeper into the production of new designed systems, we must first master computational methods, reducing the number of software (SIMION, EOD, MEBS, COMSOL, etc.) used in this optimization loop. Hence, before starting the description of the three optical solutions we would like to explore to improve the CFEG properties, I would like to describe the direction we will follow to definitively address the computational methods challenge.

3.2. Computational strategy for the design of charged-particle optics system

3.2.1 Introduction

Carrying out simulation of charged particles trajectories through optical system requires various steps, briefly introduced in Section 1.3.2.8, which can be summarized as follow:

(1) implement the field computation in the three-dimensional space located between the electrodes or the pole piece using numerical methods such as FDM, FEM or BEM, or using analytical models.

(2) either perform direct ray tracing using the mechanical method by solving numerically the Newton equations describing the fundamental law of the dynamics

(3) or determine the two paraxial fundamental rays (the marginal and the principal rays) by numerically solving the paraxial equations and, in a second time, compute the aberration coefficients of Eq. (1.3) by numerical evaluation of the aberration integrals obtained using the eikonal method.

During our previous works, we have implemented the computation of trajectories using many disparate methods:

(1) paraxial fundamental rays tracing using MATLAB software and analytical field model
(2) FDM field computation and real tracing simulations using SIMION
(3) FDM field computation using SIMION and paraxial fundamental rays tracing using python software
(4) FEM field computation and real tracing simulations using EOD
(5) FEM field computation, paraxial fundamental rays tracing and aberrations determination using EOD
(6) FEM field computation using COMSOL, paraxial fundamental rays tracing and aberrations determination using MATLAB software

None of these methods was satisfactory to implement an optimization procedure even for a simple optical system. Indeed, optimization requires the definitions of *operands* for instance paraxial ones such as the principal or marginal ray positions, or their slopes, in a specific plane. It could also be aberration values in that plane. Then, we will have to determine a set of *variables* such as the electrodes distances, the lens excitation for a single lens, and the distance between lenses inside a complete system. An ideal, and hypothetical for now, optimization code will try to fit an operand requirement set by the user (minimizing one aberration for instance) adjusting the set of selected variables. If many operands have to be optimized at the same time, we will have to give them a specific *weight* defining the priority in the optimization loop allowing to reach a realistic solution in a short time. This procedure cannot be easily automated using the previous methods due to difficulties in the definition of the variables (EOD, COMSOL) or the calculations of a paraxial or nonlinear operands (SIMION). Technically, the last option based on COMSOL FEM field computation, paraxial fundamental rays tracing, and aberrations determination within MATLAB, remains the most promising method examined for now, because an optimization code could be written in MATLAB due to its compatibility with COMSOL. However, this solution is not entirely satisfactory, because it is based on commercial software solutions which can become obsolete with time, thus maintaining our dependance to the providing company. Finally, in order to complete this task once and for all, I think the easiest way will be to develop our own internal code from the field calculation to the final optimization. We have now enough background following our previous code, to be able to reach this goal. Furthermore, a platform such as python offers many possible open-source packages for the field calculations using BEM for instance (Betcke and Scroggs, 2021) which should be easily implementable within our previous python codes. In the following, I will detail the strategy I would like to follow.

3.2.2 Field calculation

3.2.2.1 Field description in electron optical system

First of all we shall deal with the calculation of electrostatic or magnetostatic fields, i.e., time-independent fields which will be always true in our future design. Radio-frequency (RF) electrostatic systems could be nevertheless very interesting considering the optimization of the aperture and chromatic aberrations as already pointed out by Scherzer (1936), but for now we will restrict our code developments to static fields. The space through which particles travel, and where the fields have to be determined, will be referred to the "extended paraxial" domain located between electrodes and pole piece (Hawkes and Kasper, 2018). The final goal will be to derive suitable series expansions for the field within this domain. Indeed the various coefficients encountered in the field series expansions will be used in the next parts for the computation of paraxial beam properties as well as aberration coefficients. Even if the field is only needed in this so-called extended paraxial domain, it finally requires its complete calculation within the whole device. Indeed, static field within a given domain can only be calculated as the solution of a boundary-value problem. In our case, the boundaries are defined by the electrodes or pole-pieces surfaces. Boundary-values problem will be different considering a magnetostatic or electrostatic system.

The boundary value problem for the electrostatic field can be defined by each electrodes potential $U(\vec{r})$. The corresponding value for one electrode must be constant anywhere on its surface. Whereas the material properties of the metallic electrodes are not interesting in electrostatics, since the electrostatic field vanishes inside any conductor, the situation in magnetostatics is far more complicated as the magnetic field inside pole-pieces does not vanish. Unfortunately, it is not always possible to restrict the computation in the lens gap because a knowledge of the field distribution in the yoke is often needed when the shape of the latter is being designed and optimized.

Hence, lens excitation in ampere-turns will be needed to estimate the magnetic field strength $\vec{H}$. Additionally, yoke and pole pieces magnetization properties will also be required to finally determine the magnetic flux density $\vec{B}$ which is the only relevant vector field acting on the charged particle.

These fields will be extracted by computing their associated potential distribution in the extended paraxial domain considering specific boundary-value problem. This distribution is expressed thanks to partial differential equations originally extracted from the four Maxwell's equations of statics fields (Hawkes and Kasper, 2018). For an electrostatic problem considering a domain free of charges (i.e., the space charge density $\rho(\vec{r}) = 0$), the goal will be to compute the electrostatic potential anywhere using the Laplace equation:

$$\vec{\nabla}^2 U(\vec{r}) = 0 \tag{3.1}$$

while it will require to compute the magnetic vector potential $\vec{A}$ anywhere using the Poisson equation in the case of a magnetostatic field:

$$\vec{\nabla}^2 \vec{A}(\vec{r}) = -\mu \vec{j}(\vec{r}) \tag{3.2}$$

where $\vec{j}(\vec{r})$ is the electric current density in the simulation volume and μ the permeability.

Furthermore, in connection with the finite-element method, it is of importance that the partial differential equations given above can also be derived from a variational principle, the integrand being the stored field energy. The general form of variational principle for n coupled functions of position $f_i(\vec{r})$ (with $i = 1$–n) in a three-dimensional domain D is given by:

$$\delta F = \delta \int\int\int_D \Lambda(\vec{r}, f_1, f_2, \ldots, f_n, \nabla f_1, \nabla f_2 \ldots, \nabla f_n) d^3\vec{r} = 0 \tag{3.3}$$

where Λ is the corresponding Lagrange density. Least action principle described previously in Eq. (1.9) is a specific problem which uses the variational principle. In the case of an electrostatic system, we have $n = 1$ and $f_1(\vec{r}) = U(\vec{r})$. Hence, considering a domain free of charges, the Lagrange density will be written:

$$\Lambda = \frac{\epsilon}{2} \vec{\nabla}.U(\vec{r}) \vec{\nabla} U(\vec{r}) \tag{3.4}$$

where ϵ is the permittivity. Minimizing the associated variational function $F = \int_D \Lambda dv$, i.e., the energy density, is equivalent to considering the solution of the Laplace's Eq. (3.1). Indeed, the former can be simply retrieved from the Euler–Lagrange equation using Λ given in Eq. (3.4). In the case of a general magnetostatic system, we have to consider $n=3$ and $f_i(\vec{r}) = A_i(\vec{r})$ are the three components of the magnetic potential vector. Hence, the Lagrange density will be written:

$$\Lambda = \frac{1}{2\mu}\left(\vec{\nabla} \wedge \vec{A}\right)\left(\vec{\nabla} \wedge \vec{A}\right) - \vec{j}.\vec{A} \tag{3.5}$$

Before moving on to the description of the numerical computational methods, I will first describe a few extra formalisms that are required for determining electron optical system properties.

As already pointed out, the field and potentials within the extended paraxial domain of an optical element should be expressed using series expansions. The general idea is to decompose a three-dimensional field into a sequence of uncoupled two-dimensional fields. The potentials being periodic with respect to the azimuth θ in cylindrical coordinates, it suggests that we should first introduce their Fourier series expansions (Glaser, 1952). For simplicity, we will always considered a *straight optic axis* for now. Corresponding mathematical developments expressed in a curved optic axis system can be found in the literature (Hawkes and Kasper, 2018; Rose, 2009).

Considering any scalar potential (electrostatic or even magnetostatic under the scalar magnetic potential approximation), we can develop the three-dimensional potential function $U(\vec{r})$ following the series expansion:

$$U(r) = \sum_{m=0}^{\infty} \Re\left(U_m(z,s)w^m\right) \tag{3.6}$$

expressed in the Cartesian coordinate system with $w = x + iy = r.exp(i\theta)$ and $s = r^2 = ww*$.

Inserting Eq. (3.6) inside the general Poisson equation $\vec{\nabla}^2 U(\vec{r}) = -\rho(\vec{r})/\epsilon = -S(\vec{r})$, where $S(\vec{r})$ is called the source term, we first find a sequence of uncoupled differential equations:

$$\frac{\partial^2 U_m}{\partial z^2} + \frac{\partial^2 U_m}{\partial r^2} + \frac{2m+1}{r}\frac{\partial U_m}{\partial r} = -g_m \tag{3.7}$$

where g_m corresponds to the Fourier components of the series expansion performed on $S(\vec{r})$ identical to Eq. (3.6) changing $U(\vec{r})$ by $S(\vec{r})$ and U_m

by g_m. Neglecting the space-charge effect in the extended paraxial domain, we have $S(\vec{r}) = 0$ and we simply retrieve the Eq. (3.1).

Eq. (3.7) will have then to be solved (usually limiting the Fourier development to $0 < m < 5$) as a sequence of boundary-value problems (Hawkes and Kasper, 2018).

The similar method can be applied to a magnetostatic system described using the scalar magnetic potential approximation. However, regarding a real magnetostatic system, the Fourier series expansion must be realized on each component of the magnetic vector potential with the source term defined in Eq. (3.2) by the electric current density. Inserting these new series expansions in the Poisson equation (3.2) allowed us to determine a set of new differential equations for the Fourier components, such as Eq. (3.7). I will not describe them as it doesn't bring anything new for the methodology we want to implement, and the mathematics are well described in the literature (Hawkes and Kasper, 2018).

Since the extended paraxial domain is source-free $S(\vec{r}) = 0$, we shall now investigate solutions of Eq. (3.6) with source terms vanishing for sufficiently small values of r. In that case, we can show that each Fourier coefficient m defined by $U_m(z, s)$ is uniquely determined by its axial distribution, or axial harmonics, $u_m(z) = U_m(z, 0)$ following the radial series expansion:

$$U_m\left(z, r^2\right) = \sum_{n=0}^{\infty} \frac{m!}{n!(m+n)!} \left(-\frac{r^2}{4}\right)^m u_m^{(2n)}(z) \tag{3.8}$$

where $u_m^{(2n)}$ is the $(2n)^{th}$ derivative of $u_m(z)$ relatively to z. In the case of a magnetostatic system, the Fourier terms encountered inside the series expansions of the magnetic vector potential components can also be written using a radial series expansions equivalent to Eq. (3.8), assuming that the domain of solution is also source-free. All these important series expansions are well described in the literature (Hawkes and Kasper, 2018; Rose, 2009; Schwertfeger and Kasper, 1974; Sturrock, 1951). Practically, it is sufficient to truncate the power series expansions of the potentials after the terms of fourth order in x and y. The original publication of P.W. Hawkes contains explicit representations of the potential series expansions to the fourth order inside the extended paraxial domain of any optical device with a straight optic axis (Hawkes, 1965). For instance, considering an electrostatic system with rotationally symmetric field, such as the Butler lens, we have $m = 0$ in the

Fourier series. Assuming a domain free of charges $(S(\vec{r}) = 0)$, we find that the electrostatic potential will be written following the well known expansion:

$$U(r,z) = u(z) - \frac{r^2}{4}u''(z) + \frac{r^4}{64}u^{(4)}(z) - \frac{r^6}{2304}u^{(6)}(z) + \ldots \quad (3.9)$$

where $u(z) = u_0(z) = U_0(z, 0)$ is the zero-order axial harmonic simply corresponding to the familiar lens axial potential. Multipole fields series expansions can be obtained using the same method considering $m = 1$ for a dipole field, $m = 2$ for a quadrupole field, and so on (Hawkes and Kasper, 2018; Rose, 2009).

We can now carry out the field calculation once a sequence of axial harmonics $u_m(z)$, $m = 0, 1, 2, \ldots$ is known, but these functions are so far not known. They can indeed only be specified by their boundary values and by the source distribution (space charge density or electric current density). For instance, considering an electrostatic system, since all electrode surfaces must be equipotentials, the field strength on their vacuum side is given by:

$$\vec{E}(\vec{r}) = -\frac{1}{\epsilon_0}\sigma(\vec{r})\vec{n}(\vec{r}) \quad (3.10)$$

$$-\vec{E}(\vec{r}).\vec{n}(\vec{r}) = \vec{n}(\vec{r}).\vec{\nabla}U(\vec{r}) = \frac{\sigma(\vec{r})}{\epsilon_0} \quad (3.11)$$

where $\sigma(\vec{r})$ is the surface charge density and $\vec{n}(\vec{r})$ is the surface normal directed outwards from the medium (electrodes, pole pieces) (Kittel, 2005). Boundary conditions in magnetostatic systems can be found in the literature and are determined by applying Biot–Savart's Law to the electric current density condition and taking the surface magnetization distribution of ferromagnetic pole pieces into account (Hawkes and Kasper, 2018).

3.2.2.2 *Integral equations and boundary element method*

Solving a boundary-value problem can be also reduced to the solution determination of an integral equation (Riley et al., 2006). Computational methods known as boundary-element methods (BEM) used to estimate the concrete numerical solutions of these integral equations will now be discussed after a brief description of the formalism behind these integral equations. Indeed, in close collaboration with Dr. Axel Lubk from the Leibniz institute for solid state and material research in Dresden who is also looking for source-free CPO computational codes for its own instrumental developments, we would like first to implement these methods for CPO systems using existing and source-free python package such as the BEMPP package

(Betcke and Scroggs, 2021). These approaches were largely developed in Python, and there are now entire packages and a large user community accessible. In comparison to writing our own BEM, or FDM and FEM, code from scratch, these existing packages appear to be the simplest way to implement such a sophisticated computational method necessary for the three-dimensional determination of fields.

Solving a uniquely specified boundary-value problem for the Poisson's equation $\vec{\nabla}^2 U(\vec{r}) = -\rho(\vec{r})/\epsilon = -S(\vec{r})$ can be also described mathematically writing the problem in boundary integral form using the Green's theorem. It allows us to represent solutions of certain types of partial differential equations using only values and derivatives on the boundary (Riley et al., 2006). Let's consider a three-dimensional domain Γ and its boundary $\delta\Gamma$, the potential in all the domain Γ given by the Poisson's equation can be written following the expression

$$\frac{1}{4\pi}\Omega(\vec{r})U(\vec{r}) = \int_{\Gamma} G(\vec{r},\vec{r}')S(\vec{r}')d^3r' + \int_{\delta\Gamma}\left(G(\vec{r},\vec{r}')\sigma(\vec{r}') - \vec{n}'.\vec{\nabla}'G(\vec{r},\vec{r}')U(\vec{r}')\right)da' \quad (3.12)$$

where $G(\vec{r},\vec{r}') = (4\pi|\vec{r}-\vec{r}'|)^{-1}$ is the free-space Green's function, $\vec{\nabla}'$ is the derivative operator over the variable of integration $\vec{r}'$, $\Omega(\vec{r}) = 1$ for point $\vec{r}$ inside the domain Γ, $\Omega(\vec{r}) = 1/2$ for point $(\vec{r})$ inside the boundary $\delta\Gamma$, and finally $\Omega(\vec{r}) = 0$ for point $(\vec{r})$ outside the domain Γ (Hawkes and Kasper, 2018).

The first right-hand side of Eq. (3.12) is a space-source term, the second a surface-source term and the last is a surface-polarization term. Yet, the boundary functions $U(\vec{r}')$ and $\sigma(\vec{r}')$ are still independent and we have to define one additional boundary condition. Defining uniquely the boundary value $U(\vec{r}')$ while $\sigma(\vec{r}')$ remains unknown is called *Dirichlet* boundary value problem, while the opposite situation is called *Neumann* boundary value problem. Within the Dirichlet problem, the integral Eq. (3.12) becomes an integral equation of Fredholm's first kind for the boundary values of $\sigma(\vec{r}')$ written:

$$U(\vec{r}) = \int_{\Gamma} G(\vec{r},\vec{r}')S(\vec{r}')d^3r' + \int_{\delta\Gamma} G(\vec{r},\vec{r}')\sigma(\vec{r}')da' \quad (3.13)$$

After solving Eq. (3.13), we could then determine the potential at any point $\vec{r}$ in the domain Γ. Describing the problem using Neumann boundary conditions, the integral Eq. (3.12) becomes an integral equation of Fredholm's second kind for the boundary value $U(\vec{r})$, which can also be used to determine $U(\vec{r})$ at any arbitrary points inside Γ (Fredholm, 1903). Following the same methodology, similar integral equations can be also derived from the Green's theorem applied to the magnetostatic Poisson's Eq. (3.2). Various form of integral equations for a magnetostatic problems have been demonstrated depending on the situation and can be found in the literature (Hawkes and Kasper, 2018; Jagannathan and Khan, 2019; Rose, 2009).

Thus, carrying out the determination of three-dimensional unknown functions can be obtained by means of two-dimensional integral equations usually only soluble numerically using BEM methods. This reduction of a three-dimensional problem in a two-dimensional one is the main strength of these integral/BEM methods. Considering, cylindrically symmetric system we can even reduce the problem by one dimension and describe it by means of one-dimensional integral equations. Indeed, following the Fourier decomposition of the potential described by Eq. (3.6), and neglecting the source term $S(\vec{r}) = 0$ in the extended paraxial region, we find that Eq. (3.13) using Dirichlet boundary value problem can be written as a sequence of uncoupled one-dimensional Fredholm equations given by (for $m=0, 1, 2, \ldots$):

$$\nu_m(s) = \int_C G_m(\vec{u}(s)\vec{u}(s'))r(s')\sigma_m(s')ds' \tag{3.14}$$

where, considering cylindrical coordinate system with the vector $\vec{u} = (r, z), \nu_m(s)$ is defined by the inverse Fourier transform of $U(\vec{u}(s), \varphi)$ with respect to the variable φ.

Following the same logic $G_m(\vec{u}(s), \vec{u}(s'))$ are essentially the Fourier coefficients of the free-space Green's function $G(\vec{r}, \vec{r}')$ with respect to φ, and $\sigma_m(s')$ are the Fourier coefficients of the surface charge density with respect to $\varphi : \sigma(\vec{r}) = \sum_{m=0}^{\infty} \Re(\sigma_m(s)e^{im\varphi})$. $G_m(\vec{u}(s), \vec{u}(s'))$ can be evaluated analytically and the resulting expressions contain complete elliptic integrals. Then Eq. (3.14) can be also estimated thanks to numerical computation using BEM methods. Same one-dimensional integral equations can be also extracted for the Fourier components of the scalar magnetic potential or each components of the magnetic vector potential in cylindrically symmetric magnetostatic problem (Hawkes and Kasper, 2018).

We can now discuss the numerical solution of these integral equations using BEM methods. In the electrostatic case, the physical idea behind BEM method will be to compute the charge density distribution corresponding to the voltages applied to the electrodes. Finally, the potential distribution in the domain given by this charge distribution will be calculated using integral equations (Hawkes and Kasper, 2018; Kasper, 1983; Kasper and Scherle, 1982). One can start with an arbitrary guess of the charge-density distribution. Then the differences between the given and the calculated values of the electrode potentials can serve as bases to obtain improved values of the surface-charge densities and so on (using iterations). Same philosophy for magnetostatic system as described in the literature, see for instance the work in the group of Murata (Murata et al., 2004, 2016). The BEM method finds an approximation by discretizing these boundary integral equations either using one-dimensional equations in cylindrical symmetric system, or using two-dimensional ones in the most general case. The mathematical background of BEM is well covered in Steinbach (2008) and McLean (2000) and in Hawkes and Kasper (2018) and Jagannathan and Khan (2019) for the specific case of electron optics. A clear overview of the mathematical development can be also found in the thesis of Boudarham (2011). As a result, I won't go through it in depth here; instead, I will discuss how this method could be implemented using the Bempp package, which we would like to test for our field calculations (Betcke and Scroggs, 2021).

In order to solve a problem using the Bempp package, we must first define the mesh on which the problem will be discretized. In Bempp, the grid will be a triangulation of a 2D surface in 3D space and it only supports grids consisting of flat surface triangles. Once we have created a surface mesh, Bempp needs to define finite-dimensional function spaces on the grid. This is important for choosing the right type of discretization in actual computations. Indeed, in order for Eq. (3.12) to make sense, we require all terms to be finite. This motivates the definition of the following spaces:

(1) The space of square integrable functions $L^2(\Omega)$ of all functions f that satisfy $\int |f|^2 d^3r' < \infty$

(2) The space $H^1(\Omega)$ of all square integrable functions f in $L^2(\Omega)$ that also have a square integrable derivative, that is $\int \left|\vec{\nabla}.f\right|^2 d^3r' < \infty$. This space is called a Sobolev space.

The space $H^1(\Omega)$ can also be interpreted as a space of functions with bounded energy. It therefore makes physical sense to look for solutions in this space, and indeed it is the right space of functions for Poisson's problem. Other important boundary spaces are fractional Sobolev spaces $H^{1/2}(\Omega)$ and $H^{-1/2}(\Omega)$. There are different ways of defining fractional Sobolev spaces. But this is a technical topic, which I will not cover in more detail here. Complete description is reported in the Bempp package webpage as well as in the following papers (Betcke and Scroggs, 2021; Scroggs et al.,

2017). These spaces will then be used to discretize the boundary integral formulations of our problem. In our case, solving electrostatic problems or magnetostatic problems will required the use of scalar function spaces. "Discontinuous polynomials," "continuous polynomials," and "dual spaces on the barycentrically refined grid" functions space are supported by the Bempp package. Solving more complicate problems using full Maxwell's equation, will require vector function spaces. Bempp supports various kind of vector function spaces such as "Rao–Wilson–Glisson functions space" (RWG), "Scaled Nédélec functions space" (SNC), "Buffa–Christiansen functions space" (BC), or "Rotated Buffa–Christiansen functions space" (RBC). All the mathematical details can be found in Scroggs et al. (2017) as well as on the Bempp webpage. More details and references can also be found in the review articles (Buffa and Hiptmair, 2003) and in Meury (2007). In order to set the main concept behind function space down, I have reported the description of two scalar functions spaces used in Bempp in Fig. 3.1A and B.

With the definition of a function space, we can now discretize functions over a given space into a grid function object. Indeed, in Bempp, data on a given grid is represented as a grid function object. A grid function consists of a set of basis function coefficients and a corresponding function space (Scroggs et al., 2017). We can now define an operator acting on grid functions either using Dirichlet, Neumann, or mixed boundary conditions. Boundary integral formulations of these problems, are commonly written in BEM using boundary integral operators. They simply correspond to the potential operators, given formally by the Eq. (3.12) and described in the following, taken at the boundary limit $\vec{r} \to \delta\Gamma$. When creating a boundary integral operator in Bempp, three spaces are provided: the domain, the range, and the dual to the range (given as inputs in that order). The domain and dual spaces are used to calculate the weak form of the operator. The range is used by the operator algebra to assemble operator products correctly (Scroggs et al., 2017). Discretizing the identity operator leads to a matrix $M = (m_{ij})$ with the following general form:

$$m_{ij} = \int_{\delta\Gamma} \phi_i(x)\psi_j(y)dxdy \tag{3.15}$$

where $\phi_i(x)$, $\psi_j(y)$ are the basis functions of the domain and dual spaces, respectively. For Laplace's equation used in electrostatic problem, there are four boundary operators that are used in Bempp which are reported in Scroggs et al. (2017). For instance, the simplest one is the so-called single

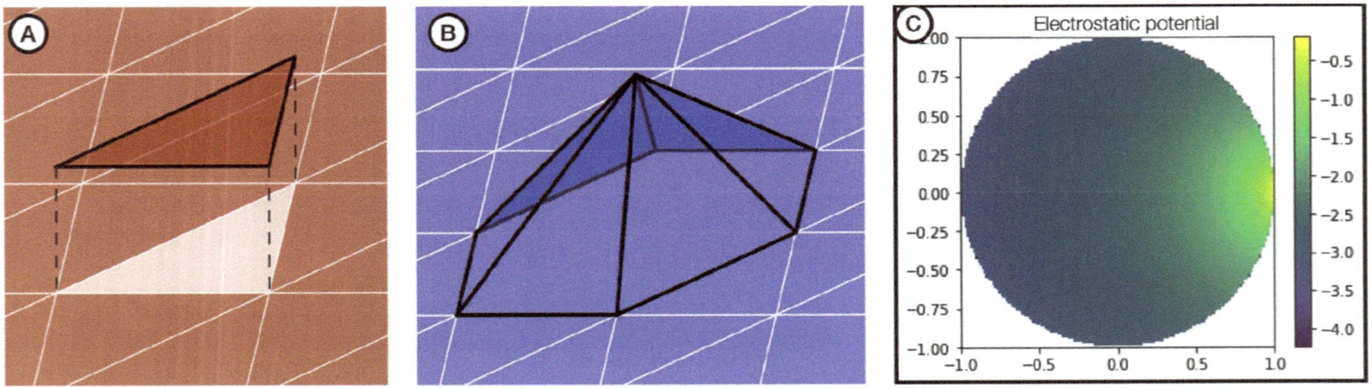

Fig. 3.1 (A) Discontinuous polynomial order 0 basis function. (B) Continuous polynomial order 1 basis function. From the Bempp webpage (see https://bempp.com). (C) Electrostatic potential inside a unit sphere determined using the Bempp package considering Dirichlet boundary conditions (source located at the point (0.9, 0, 0)).

layer operator will have the classical form encountered in the integral equations:

$$m_{ij} = \int_{\delta\Gamma} G(x, y)\phi_i(x)\psi_j(y)\,dxdy \tag{3.16}$$

where $G(x, y)$ is the free-space Green's function. Similar operators will be generated for Poisson's equation problem. It is worth noting that Bempp was created to compute more complicate electromagnetic problems requiring the use of Maxwell's equations instead of the Laplace or even the Poisson equation. Other operators can be also generated in Bempp using Green's function for the Helmholtz equation depending on the wavenumber (Scroggs et al., 2017). In our case, the situation is simpler as we will always consider static problems. Once we have assembled the relevant operators, and have created a grid function containing the relevant right-hand-side data, we will need to solve our linear system. The computation can be performed using a direct solver, or an iterative solver. Direct solvers compute the solution of a linear system by computing the inverse of the matrix. Iterative solvers solve a linear system iteratively. Finally, once the solution of a boundary integral formulation has been approximated, potential operators can be used to compute point evaluations of the solution inside the domain. For instance, considering Laplace's equation, there are two potential operators $SL(f)$ and $DL(f)$ that are used in Bempp, given by the Eq. (3.12), which are in one-dimension:

$$SL(f)(x) = \int_{\delta\Gamma} G(x, y)f(y)\,dy \tag{3.17}$$

$$DL(f)(x) = \int_{\delta\Gamma} \frac{\partial G(x, y)}{\partial f_y} f(y)\,dy \tag{3.18}$$

To assemble potential operators in Bempp, first the desired evaluation points must be defined and then we use Green's representation theorem to evaluate the solution on these points. The solution can finally be plotted using python. Fig. 3.1C shows the solution of a simple Laplace problem performed following the on-line tutorial (http://bempp.com/documentation/tutorials.html). The computation solves the electrostatic potential inside a unit sphere with Dirichlet boundary conditions (i.e., unknown Neumann conditions). The boundary data is a source located at the point (0.9, 0, 0) (see the online page for the boundary source equation).

We can then see that Bempp is well suited for the implementation of an electrostatic problem. Bempp offers also the possibility to solve the problem considering Dirichlet and Neumann boundary conditions applied to different parts of the domain boundary. Additionally, it allows BEM and FEM to be coupled to solve one specific problem (Betcke and Scroggs, 2021). The implementation of magnetostatic problem considering either the scalar magnetic potential or the magnetic vector potential, or even superimposed electrostatic–magnetostatic problems also seems feasible with the Bempp package by incorporating Maxwell's equation. Thus, the next step will be to use this package to map the various Fourier components of the integral equation (3.14) in order to evaluate the field series expansions in any electrostatic or magnetostatic electron optical system. In the following sections, we will see how these data will be used to determine the linear and non-linear properties of the optical system. Before going on to the computation of the optical properties, I will briefly describe how FEM can be used to solve an electrostatic–magnetostatic problem if we want to use these methods in specific situation in combination with BEM.

3.2.2.3 First-order finite element modeling (FOFEM)

As noted previously, FEM approaches can be used to solve electrostatic and magnetostatic problems. Carrying out these simulations is possible using Comsol multiphysics, MEBS or EOD (Chu and Munro, 1982; COMSOL AB, n.d.; Lencová and Zlámal, 2008). However, within the framework of my research project the goal will be to control the computation of the fields, allowing us to implement optimization procedure within an open source homemade code. This is almost impossible by combining many third-party softwares as we have experienced over the past few years. The use of an open source python package, such as Bempp, within a more general homemade python optimization code seems to be the only viable solution to this situation. However, some problems will not be easily treated using BEM methods and will require them to be combined with FEM methods (also available in Bempp as already pointed out).

One of the simplest and versatile methods is first-order finite-element modeling or FOFEM (Chu and Munro, 1982; Zienkiewicz and Morgan, 2006). For instance, solving an electrostatic problem using FOFEM requires creation of a mesh throughout the volume Γ surrounded by the boundary $\delta\Gamma$ on which the potential distribution is assumed to be known. In FOFEM, the mesh will subdivide the domain inside and between electrodes in quadrilateral volumes with different sizes following the real electrode geometry.

Finally, each fine quadrilateral is divided into two triangles called finite elements. An electrostatic potential U_i is assigned to each mesh point. The first order approximation of FOFEM will assume that the potential varies linearly with position inside each finite element. In FOFEM, the potential inside an element is then given by the potentials at its nodes and the potential distribution within the domain is finally determined by the set of potentials U_i determined for N mesh points (with $i=1, N$).

The potential is computed by minimizing the variational functional F given by Eq. (3.3) using the Lagrange density from Eq. (3.4) for an electrostatic problem or the one in Eq. (3.5) for a magnetostatic problem. Minimizing numerically, the variational functional requires assignation of a potential U_i to each mesh points. The functional F is expressed as a quadratic function of the mesh-point potentials. Solving $\delta F=0$ will then generate a set of finite element equations, each of them expressing the potential on a mesh point in terms of the eight surrounding points. These equations can be solved using different numerical methods such as Gaussian elimination, or the incomplete Cholesky conjugate gradient (ICCG) as described in detail by Munro (Chu and Munro, 1982). This finally allows us to compute the potential at each mesh point. It is then possible to plot out the equipotential lines and, for instance, the axial potential distribution $u(z)$ used in the field expansions (3.9) for a rotationally symmetric element.

I will not enter more deeply in the mathematical description of these methods, described in detail in the literature. The next step will be to find the best way to implement them either in combination with BEM or alone depending on the particular problem to be solved. For instance, the FEM is definitively a best choice when saturation effects become important in magnetostatic problems considering the ferromagnetic properties of the pole-pieces or the yoke. However, this corresponds to a very specific situation, and most of the time the BEM should be sufficient, thanks to the use of a modern python package such as Bempp.

It is worth noting that, for some applications, higher accuracy FEM can be implemented called second-order FEM or SOFEM. The main difference with FOFEM is that instead of using triangular elements with three nodes and a linear variation of potential, a second-order isoparamatric finite element with nine nodes in which the potential is a quadratic function of position, will be used. SOFEM could be advantageous for complicated geometry such as the extractor electrode of a CFEG. All details can be found in the literature (Zhu, 1989; Zhu and Munro, 1995).

3.2.3 Paraxial properties

Determining optical properties (paraxial and aberrations) of an optical system can be performed using methods based on Hamilton's characteristic function, originally proposed for conventional optics system (Gross, 2005). This is the core of the "optical approach," which was first discussed in Chapter 1. In electron optics, the Hamilton's point characteristic function is given by Eq. (1.12) and the refractive index by Eq. (1.13). It is convenient to transform the refractive index n from Eq. (1.13) into the following form (Hawkes, 1965):

$$m = \sqrt{U(\vec{r})\left(1 + \frac{e}{2m_0c^2}U(\vec{r})\right)(1 + x'^2 + y'^2)} - \sqrt{\left(\frac{e}{2m_0}\right)}\left(A_x.x' + A_y.y' + A_z\right) \quad (3.19)$$

where z is chosen to be the straight optic axis of our system. As a consequence, $n.ds$ encountered in the original Hamilton's principle reported in Eq. (1.12), can be replaced by $m.dz$. Following Eqs. (3.6) and (3.8), all the potentials $U(\vec{r})$ and $\vec{A}(\vec{r})$ can be expanded as power series in the off-axial coordinates. Inserting these expansions into m and expanding the square roots as power series also, we find that m is composed of a series of terms which we group according to their degree in coordinates x, y and their derivatives. Thus:

$$m = m^{(0)} + m^{(2)} + m^{(4)} + \ldots + m^{(2n)} + \ldots \quad (3.20)$$

in which $m^{(2n)}$ consists of all the terms of degree $2n$ in the power expansion. Only even terms appears as we will now consider system with two axes of symmetry such as round, quadrupoles, or octopoles lenses. Hexapole or magnetic/electrostatic sectors defined by only one axis of symmetry, will have also odd terms such as $m^{(3)}$ generating second-order transverse aberrations (Rose, 2009). We will restrict our description to systems with two axes of symmetry because they will be dominant in our future designs. Nevertheless, the following method could be directly applied to handle other systems, simply by considering odd order point characteristic function in the determination of aberrations.

Using the refractive index expansion (3.20), we can derive an identical expansion for the Hamilton's point characteristic function $S = S^{(0)} + S^{(2)} + S^{(4)} + \cdots$. Considering the action of a refracting surface in conventional optics, similar power expansion is obtained for the optical length V given by Eq. (1.4), i.e., the point characteristic function of a light ray (Gross, 2005). In that case, the zero-order corresponds to the optical length of the optic axis from the object plane to the image plane, the second-order to the paraxial contribution on the total optical length, fourth-order term to the primary Seidel aberrations contribution on the total optical length.

Same logic will be used in our case. Here, the paraxial equations will be extracted by applying Hamilton's principle, given by Eq. (1.12), to the second-order point characteristic function as follow:

$$\delta \int m^{(2)} dz = 0 \tag{3.21}$$

The explicit dependence of $m^{(0)}$, $m^{(2)}$, $m^{(4)}$, ... upon the axial harmonics $u_m(z)$, extracted from the Fourier coefficients U_m of the potential, and their derivatives relative to z, can be determined by inserting the series expansion given by Eqs. (3.6) and (3.8) into the refractive index given by the Eq. (3.19). In order to give an explicit representation of the zero-order and second-order terms of the refractive index, we must first introduce a more convenient expression for the axial harmonics. For instance, regarding an electrostatic system we will rewrite the axial harmonics in Eq. (3.8) as follow:

$$u_m(z) = \frac{(-1)^m}{m!}(p_m(z) - iq_m(z)) \tag{3.22}$$

where $p_m(z)$ and $q_m(z)$ are real axial harmonics which have physical meanings. For instance, $p_0(z) = u_0(z) = u(z)$ is the classical axial potential, $p_1(z)$ and $q_1(z)$ are encountered in dipole system ($m = 1$) and correspond to the transverse components of the field strength $\vec{E}$ on the optic axis along x and y, respectively, $p_2(z)$ and $q_2(z)$ functions describe the potential distribution in symmetrical electrostatic quadrupoles ($m - 2$), and so on. As usual, the same operation can be applied to magnetostatic system (Hawkes and Kasper, 2018). Hence, considering a general electrostatic system and neglecting dipole $m = 1$ terms, we can now explicitly express the zero-order and second-order order terms of the refractive index expansion as a function of the axial harmonics of the potential as follow (Hawkes, 1965):

$$m^{(0)} = \sqrt{u(z)(1 + a.u(z))} \tag{3.23}$$

$$m^{(2)} = \frac{1+2.a.u(z)}{\sqrt{u(z)(1+a.u(z))}}\left(-u''(z)\frac{1}{8}\left(x^2+y^2\right)+p_2(z)\frac{1}{8}\left(x^2-y^2\right)\right.$$
$$\left.+q_2(z)\frac{1}{2}xy\right)+\frac{1}{2}\sqrt{u(z)(1+a.u(z))(x'^2+y'^2)} \tag{3.24}$$

where the notation $a = e/(2m_0c^2)$ is used to simplify the mathematical expressions.

Equivalent expressions of second-order point characteristic function $S^{(2)} = \int m^{(2)} dz$ can be derived for magnetostatic system, or even mixed electrostatic–magnetostatic, by simply considering the series expansion of each magnetic vector potential components in the calculation of the refractive index expansion. The Cartesian coordinate system x, y used in the previous formula is fixed in the laboratory frame. Expressions in the rotating coordinate system are important especially for magnetostatic system. They can be derived very easily from previous equations and they can be found in the literature (Hawkes, 1965; Hawkes and Kasper, 2018).

Carrying out minimization of point characteristic function using Eq. (3.21), or solving Euler–Lagrange differential equations considering $m^{(2)}$ as the Lagrangian function which is strictly equivalent, will give the paraxial equations. This method is general, and can be applied to electrostatic, magnetostatic, or mixed systems with rotationally or quadrupole symmetry. It is obvious from Eq. (3.24) that octopoles will not have any paraxial effects (only $m = 2$ multipole contribution appears coming from quadrupole paraxial action, no higher order). For instance, limiting the $m^{(2)}$ development to zero-order magnetic axial harmonics (corresponding to the action of the rotationally symmetric elements) we can derive the simple paraxial Eq. (1.60) of round magnetic system from the Hamilton's principle. The same procedure is used to extract paraxial equation of round electrostatic lenses given by (Chu and Munro, 1982):

$$r''(z) + \frac{u'(z)}{2u(z)}r'(z) + \frac{u''(z)}{4u(z)}r(z) = 0 \tag{3.25}$$

or the paraxial equations of quadrupoles lenses by retaining the $m = 2$ axial harmonics (Hawkes, 1965). In fact, quadrupole-based solutions will be also examined within my project for a new aberration-corrected electron gun concept which will be described in the following sections. As a consequence, implementing this general point characteristic-based method will be well adapted to determine the paraxial properties of any kind of systems addressed in the future project.

Solving the paraxial Eq. (3.21) will enable us to determine the two fundamental rays, i.e., the principal ray $\vec{g}(z)$ and the marginal ray $\vec{h}(z)$ previously introduced in Chapter 1. In the fixed Cartesian coordinate system x, y, z, the general solutions of any paraxial trajectory will then be formally written:

$$\begin{aligned} x(z) &= x_o g_x(z) + x_p h_x(z) \\ y(z) &= y_o g_y(z) + y_p h_y(z) \end{aligned} \tag{3.26}$$

where x_o, y_o are the coordinates of the starting point of the trajectory in the object plane, and x_p, y_p the positions in the pupil plane or the aperture stop plane. x_p, y_p can be expressed as a function of a_0 and b_0, originally used in expressions (1.12) and (1.3). In CPO, we will usually choose the two fundamental rays which fulfill the boundary conditions $g(z_o)=h(z_p)=1$ and $g(z_p)=h(z_0)=0$.

Paraxial equations can be solved numerically using fourth-order Runge–Kutta formula (Kutta, 1901). A python software has been already realized during the Julien Dupuy's thesis described in Section 1.3.2.8. This code allows to compute paraxial solutions of round magnetic lenses taken the discretized magnetic field values determined by SIMION thanks to its FDM approach. Within the framework of my project, this code will need to be modified to solve Eq. (3.21) for any type of system (electrostatic or magnetostatic, round or quadrupole lenses) using the appropriate second-order term in the refractive index expansion. Axial harmonics will have to be extracted from the field calculation performed within the same code, for instance, using the Bempp package. This part should be straightforward if the field computation can be properly implemented.

3.2.4 Aberrations properties

Using the Hamilton's point characteristic function expansion derived from Eq. (3.20), we can now determine the contribution of each primary Seidel aberration. Primary aberrations are derived by considering the fourth-order point characteristic function $S^{(4)}$ (Hawkes, 1965). Such as for second-order transverse geometrical aberration terms in Eq. (1.3), third-order point characteristic function vanishes in system with two planes of symmetry. They will then have to be considered only in systems with one plane of symmetry such as hexapoles or sectors.

The fourth-order point characteristic function can be written:

$$S^{(4)} = \int m^{(4)} dz \tag{3.27}$$

where the fourth-order term of the refractive index expansion have to be written as a function of object and pupil coordinates after substituting paraxial solutions (3.26) in the expression of $m^{(4)}$. Such as for the $m^{(2)}$ term in (3.24), $m^{(4)}$ will have first to be expressed using axial harmonics. Details of these long calculations steps can be found in the impressive series of papers and books by P.W. Hawkes for instance (Hawkes, 1965; Hawkes and Kasper, 2018). Finally, the fourth-order refractive index can be written in the following form:

$$\begin{aligned} m^{(4)} &= \sum_{k,l,m,n} a_{klmn} x_p^k y_p^l x_0^m y_0^n \\ &= a_{0040} x_0^4 + a_{0004} y_0^4 + a_{4000} x_p^4 + a_{0400} y_p^4 + a_{0031} x_0^3 y_0 + \ldots \end{aligned} \tag{3.28}$$

where a_{klmn} can be expressed using real axial harmonics $p_m(z)$ and $q_m(z)$ and their derivatives relative to z (see Hawkes, 1965; Hawkes and Kasper, 2018 and also Septier, 1967).

Hamilton's point characteristic function S is a function of four variables. These variables are the object x_o, y_o and image x_i, y_i positions given by the integration limits z_o and z_i chosen to perform the integration of the refractive index in (1.12). Let's note it $S_{oi}^p = f(x_0, y_0, x_i, y_i)$ from here, with p is a reminder that this is the "point" characteristic function. The derivative of S_{oi}^p relative to any variables x_o, y_o, x_i, y_i will provide the coordinates, along x, y, of the canonical momentum vector $\vec{p}$ either in the object z_o or the image z_i areas, depending on the variables chosen to perform the differentiation (see Eq. (1.6)).

Apart from the point characteristic function, it is worth noting that other characteristic functions can be derived from S_{oi}^p by Legendre transformation, which replaces one set of variables by another (Gross, 2005; Rose, 2009). Indeed, starting from Eq. (1.12), it is possible to generate the point-angular (or mixed) characteristic function $S_{oi}^{pa} = TL_{\vec{r}_i/\vec{p}_i}(S_{oi}^p) = f\left(x_o, y_o, p_{xi}, p_{yi}\right)$ where $TL_{\vec{r}_i/\vec{p}_i}$ is the Legendre transformation that replaces point coordinates $\vec{r}_i$ by the canonical momentum ones $\vec{p}_i$ in the image space (the exponent "pa" stands for "point-angular"). Hence, we have also the angular-point (or mixed) characteristic function $S_{oi}^{ap} = TL_{\vec{r}_o/\vec{p}_o}(S_{oi}^{p}) = f\left(p_{xo}, p_{yo}, x_i, y_i\right)$ and finally the angular one $S_{oi}^{a} = TL_{\vec{r}_o/\vec{p}_o}(S_{oi}^{pa})$. The fourth-order angular characteristic function S_{oi}^{4a}, extracted from the series expansion of this function, is generally used to determine the primary Seidel transverse aberrations. It allows us to determine all third-order terms in the expansion $x(z)_{ab}$ in Eq. (1.3). Indeed, the differentiation of S_{oi}^{a} with respect to the coordinates of the momentum vector, gives the position coordinates (object side or image side depending on the choice of the variable of differentiation). Determining transverse aberrations coefficients using this method is usually implemented in conventional optics (Gross, 2005). It has been described for charged particles optics by Rose (1987, 2009).

Following the pioneering work of Sturrock, further developed in the work of Hawkes, transverse aberrations coefficients $(\ldots|\ldots)$ in Eq. (1.3) can be also extracted simply using the high-order point characteristic function, without having to go through the angular characteristic function (Hawkes and Kasper, 2018; Sturrock, 1951). The Sturrock idea was to consider higher order terms of the refraction index as a perturbation of the paraxial solutions extracted from the second-order term $m^{(2)}$. This general method, called the eikonal method, will be implemented in our case limiting the perturbation term to the fourth-order $m^{(4)}$ corresponding to primary Seidel aberrations (see Eq. (3.28)). Of course, secondary aberrations will have to be considered if the primary aberrations vanish. In that case, this perturbation method will have to be modified to take into account the sixth-order term in the refraction index expansion.

Using the eikonal method, we can now derive the transverse aberrations expansion, given by Eq. (1.3) for the x coordinate, in terms of the derivative of the fourth-order point characteristic function and the two paraxial solutions. Considering transverse aberrations along the two directions x and y, Sturrock's formula can be written as follow:

$$x(z_i)_{ab} = \frac{h_x(z_i)\left(\partial S_{pi}^{4p}/\partial x_o\right) - g_x(z_i)\left(\partial S_{oi}^{4p}/\partial x_p\right)}{\sqrt{u(z_i)(1 + a.u(z_i))}\left(g_x(z_i)h_x'(z_i) - g_x'(z_i)h_x(z_i)\right)} \tag{3.29}$$

$$y(z_i)_{ab} = \frac{h_y(z_i)\left(\partial S_{pi}^{4p}/\partial y_o\right) - g_y(z_i)\left(\partial S_{oi}^{4p}/\partial y_p\right)}{\sqrt{u(z_i)(1 + a.u(z_i))}\left(g_y(z_i)h_y'(z_i) - g_y'(z_i)h_y(z_i)\right)} \tag{3.30}$$

These expressions allow us to determine each transverse aberration coefficient in an arbitrary plane located in the image side at a position z_i. S_{pi}^{4p} is the fourth-order point characteristic function calculated by integrating the fourth-order refractive index between the pupil plane (usually the aperture stop) located in z_p and the final plane z_i: $S_{pi}^{4p} = \int_{z_p}^{z_i} m^{(4)} dz$. z_i corresponds to a random plane, not necessarily the Gaussian image plane.

Thus knowing the paraxial solutions and the fourth-order point characteristic function, we can determine any aberration coefficients for any kind of system. The fourth-order point characteristic function S^{4p} is given by integral formula. Indeed using Eq. (3.28), we have:

$$S^{(4)} = \sum_{k,l,m,n} \left(x_p^k y_p^l x_0^m y_0^n \int a_{klmn} dz \right) \tag{3.31}$$

Now that all the formalism has been properly described, we can propose an optimum coding strategy allowing us to determine the optical properties of any optical system.

So, what data are we looking for and how can we get it using numerical computation? First, we must determine all the axial harmonics $p_m(z)$ and $q_m(z)$ in the field series expansions of our system, for instance, using BEM methods. These functions appears in the refractive index expansion. Each component $m^{(0)}$, $m^{(2)}$ and $m^{(4)}$ can then be determined as a function of z. The paraxial equations can be solved using $m^{(2)}$ and the two fundamental functions $g(z)$ and $h(z)$ must be extracted taking care of their boundary conditions. Inside $m^{(4)}$ all functions a_{klmn} must be also expressed using the axial harmonics given by the BEM calculations, and the two paraxial solutions. To determine S_{oi}^{4p} and S_{pi}^{4p}, the integration of each a_{klmn} functions can be performed numerically using Simpson's rule easily implemented in python (Press, 1988). Combining the paraxial solutions and the fourth-order point characteristic function, estimated in the previous steps, inside Eq. (3.29), we can compute transverse aberrations $x(z_i)_{ab}$ and $y(z_i)_{ab}$. Each coefficients $(\ldots|\ldots)$ will be extracted looking at their power dependance in x_0, y_0, x_p/a_o, y_p/b_o, i.e., considering the well-known aberration integral $\int a_{klmn} dz$ in the development (3.31).

During the first work of Yudai Kubo performed in collaboration with HHT, followed by the PhD thesis of Julien Dupuy, we have implemented real tracing simulations of a complete instrument using only one third party software, namely, SIMION (see Section 1.3.2.8). Transverse aberration coefficients cannot be directly determined by real tracing as we don't have access to the expansion terms of the refractive index, but the Lua capabilities of SIMION gave us hope that they could be extracted and visualized. During the internship of Yann Tavernier, we succeed in computing numerically the point characteristic function S^p_{oi} for each trajectory inside SIMION, thanks to a homemade LUA code. Carrying out paraxial calculations, we could also extract all cardinal elements (image, entrance and exit pupil planes, aperture-stop plane) (see Fig. 3.2A). Finally, this code computes the function S_{op} for all rays, by numerically summing the refractive index m determined in each grid unit of the SIMION FDM mesh, as well as the one given by the paraxial principal ray (from the object plane z_o to the exit pupil plane z_p). By subtracting each ray point characteristic function from the principal one, we were able to map, for one object point taken in the field, the wavefront aberration as reported in Fig. 3.2B. Finally, each aberration order can be estimated by fitting the wavefront aberration surface using Zernike polynomials (see Fig. 1.6). This operation must be then performed for many field points to have access to off-axial aberrations.

This approach was originally used to improve the original SIMION code, which was one of my key goals following Julien Dupuy's PhD (see Section 1.3.2.8). Unfortunately, using this method mapping the wavefront aberration of one of the simplest lens geometry reported in Fig. 3.2B, took almost 30 min with a modern computer. This is why I came to the conclusion that SIMION was not well suited to optimize a complete optical

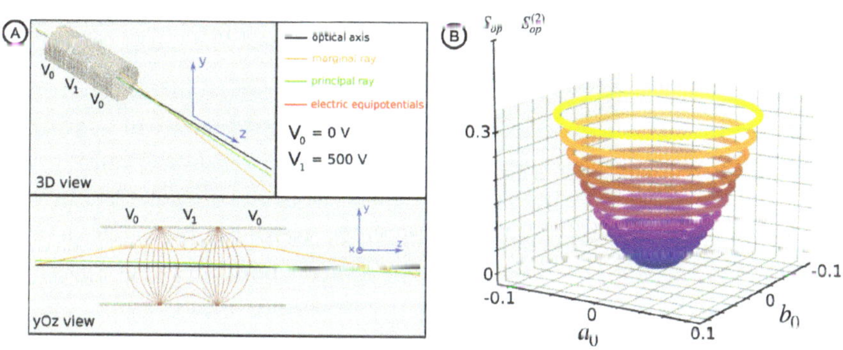

Fig. 3.2 (A) Paraxial fundamental rays of a simple electrostatic Einzel round lens. (B) Wavefront aberration mapped in the exit pupil plane of the Einzel lens, computed within SIMION thanks to a homemade LUA code. The estimation is performed by subtracting the point characteristic function S_{op} extracted for each ray, starting from one field position x_0, y_0 (0, 0 in that case) with starting angles a_0, b_0 (or pupil coordinates x_p, y_p), with the point characteristic function $S^{(2)}_{op}$ of the principal ray.

system. Several aberration estimations would be necessary and, to be realistic, they should be computed in less than a second.

Therefore, my strategy will be to determine system aberrations within the same code as the one used for field and paraxial computation, by implementing numerical methods to estimate the integral formula derived from the eikonal approach previously introduced. This strategy has been first tested and validated on a round Einzel lens mixing COMSOL and MATLAB, but the goal will be to transfer it to an open source language such as python with full control of the field estimation. This should allowed us to implement optical system optimization methods more effectively.

The chromatic aberration may be treated in a similar fashion. For the transverse primary axial chromatic aberration $(x|a\delta)$, the perturbation point characteristic function $S^{(cp)}$ which must be introduced in Eq. (3.29) (corresponding to $S^{(4p)}$ in the case of the primary geometrical aberrations) is given by $(\partial S^{(2p)}/\partial U)\Delta U + (\partial S^{(2p)}/\partial B)\Delta B$ where U, B are the electrostatic potential and the magnetic flux density, respectively. In practice, if we consider a magnetic system we take ΔU to be a variation of the acceleration voltage V_0 and ΔB a variation of the maximum magnetic field B_0. The final expression of the chromatic perturbation term $S^{(cp)}$ as a function of U, B, ΔU, and ΔB derived from this method can be found in the literature (Hawkes and Kasper, 2018).

3.2.5 System design optimization strategy

The overall strategy for the optical system optimization will be identical with classical methods used in such software as OSLO (Lambda Research Corporation, 2021, n.d.). The term optimization refers to the improvement of the performance of an optical system by changing the values of a subset of the system's constructional parameters that we called *variables*. Typically, the variables will be quantities such as bore radius, electrodes distances, excitation, etc. The system's performance will have to be measured by a user-defined error function. In OSLO, they use the weighted sum of squares of *operands*, that represents an estimate of the difference in performance between a given optical system and a system that meets all the design requirements.

The conventional approach to optimization is iterative, in which the user chooses initial values for the variables (a starting point) and an optimization algorithm is applied that repeatedly attempts to find new values for the variables that yield ever lower error function values. This approach to optimization, illustrated in the Fig. 3.3, will depend strongly upon the choice of the starting point (see Fig. 3.3A). For instance, if a starting point is chosen from

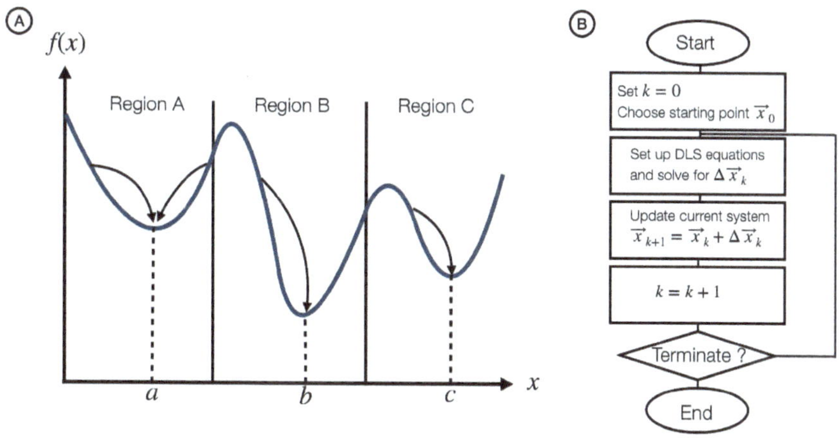

Fig. 3.3 (A) Scheme showing the importance of the starting point in the optimization process trying to reach a global minimum. (B) Damped least square optimization general algorithm used in most of modern optical software (Lambda Research Corporation, 2021).

region A or region C, the program will find the corresponding local minimum at $x = a$ or $x = c$, rather than the global minimum at $x = b$ (Lambda Research Corporation, 2021).

If the starting point is chosen from region B, however, the program will proceed to the global minimum at $x = b$. In an optical design (even rather simple), the dimensionality, i.e., corresponding to the number of independent variables, will be high and the number of local minima will be typically large, making the choice of starting point crucial to the success of optimization. In OSLO, the same starting point issues occurs, and the one from which the optimization can proceed to reach a global minimum (or a suitable local minimum) is in general be determined by experience or by finding an existing design with properties similar to those desired (Lambda Research Corporation, 2021). The importance of the choice in the starting point is also well explained in the remarkable series of books edited by H. Gross dealing with optical optimization of conventional system (Gross, 2005). Furthermore, the prime source of difficulty in optimizing a given error function is that the minimum usually depends on balancing, rather than removing, the aberrations of the system.

There are many local optimization methods available; most of those employed in optical design commercial software are *least-squares methods*, meaning that they produce a solution that is as close as possible to the desired solution when a complete solution is not possible (ordinarily there are more operands than variables). For instance in OSLO, the standard optimization algorithm is called *damped least squares* that I will briefly describe in the following.

The damped least squares (DLS) optimization is the most popular optimization algorithm in optical design softwares (Ansys Company, n.d.; Lambda Research Corporation, n.d.). In DLS, the error function $E(\vec{x})$ is expressed in the form of a weighted sum of squares:

$$E(\vec{x}) = \sum_{i=1}^{m} w_i f_i^2(\vec{x}) \tag{3.32}$$

where the vector $\vec{x} = (x_1, x_2, \ldots, x_n)$ represents the set of optimization variables and the f_i are the so-called operands. Each individual operand is defined by:

$$f_i = c_{1,i} \oplus c_{2,i} \tag{3.33}$$

The terms c are called *components*. Each operand contains two components that are linked by a math operator $\oplus$, which may be addition, subtraction, multiplication, division, exponentiation, greater than, or less than. Usually in OSLO, the first component is a ray displacement, the second component is a target value, and the operator is subtraction, but the definition (3.33) allows for more complex operands. Any quantity that can be specified as an optimization variable can also be used as a system operand component. Components c could be for instance:

(1) Aberration and paraxial data components extracted from the home-made code: primary or even secondary transverse aberrations, paraxial principal or marginal rays heights and slopes, chromatic aberrations.

(2) Ray components corresponding to values derived from aiming exact rays from specified field points at specified pupil points and tracing the rays through the system derived from real tracing software such as SIMION.

(3) Spot diagram components such as modulation transfer function (MTF), Strehl ratio or root mean squared (RMS) wavefront error, computed by tracing a spot diagram from specified field points.

Thus, the optimization code will mainly use the first type of components extracted from the homemade software. However, the code will also have the possibility to call specific ray components or spot diagram components from SIMION if the model has to be further refined. Given that the coarse time-consuming optimization will have been done using the fast homemade code, the use of third-party software at the end of the optimization process does not appear to be aberrant.

The quantities w_i in Eq. (3.33) are called *weights*, which can be used to account for the relative importance of different operands. The error function has its minimum value when all the operands are zero. It is useful to write $E(\vec{x})$ in vector notation as:

$$E(\vec{x}) = \vec{f}^T\vec{f} \tag{3.34}$$

where the matrix $\vec{f}^T$ is the transpose of $\vec{f} = (f_1, f_2, \ldots f_m)$.

Minimization of the error function is based on a piece-wise linear model of the operand dependencies on the variables. That is, the change in the ith operand due to a change in the jth variable is assumed to be given by:

$$f_i(x_j + \Delta x_j) = f(x_i) + \frac{\partial f_i}{\partial x_j}\Delta x_j \tag{3.35}$$

Of course, this is an idealized model. In a real situation, there may be nonlinearities that require second or higher-order derivatives of the operands in the above equation.

To describe the optimization of systems in which there are several operands and several variables, it is best to use matrix notation: $\bar{\bar{A}}.\Delta\vec{x} = -\vec{f}$, where $\bar{\bar{A}}$ is the matrix of the derivatives of each of the operands with respect to each of the variables ($A_{ij} = \partial f_i/\partial x_j$). Ordinarily, there are more equations than variables, so there is no direct solution. However, there is a least squares solution for which the error function has a minimum value. The least squares algorithm operates as follows: given an initial estimate $\vec{x}_0$ of the minimum (a starting point), iteratively determine new estimates of the minimum $\vec{x}_{k+1} = \vec{x}_k + \Delta\vec{x}_k$ by solving the following linear system, called the "least-squares normal equations," for the change vector $\Delta\vec{x}_k$:

$$\bar{\bar{A}}_k^T.\bar{\bar{A}}_k\Delta\vec{x}_k = -\bar{\bar{A}}_k^T.\vec{f}_k \tag{3.36}$$

where k is the iteration number. In OSLO, the above scheme sometimes fails owing to nonlinearities in the system where the change vector $\Delta\vec{x}$ typically diverges. The solution to this problem is to add a damping term μ to Eq. (3.36) that prevents large values for $\Delta\vec{x}$. The new equations are called damped least squares normal equations. I will not go further into details regarding the mathematical background which can be found in the literature (Lambda Research Corporation, 2021). The basic scheme for damped least squares iteration which we will have to implement in our final code is shown in the Fig. 3.3B.

3.2.6 Discussion

The first priority of the research project will be to implement all these methods and test the final code on well-known systems such as the Butler electrostatic lens in CFEG or any round magnetic lenses used in the I2TEM. Thanks to this robust homemade software, we will be able to study new innovative optical designs for the CFEG and optimize them before constructing them. In the next sections, I propose to work on three unique ideas, all of which appear to be quite promising for future coherent EM technology. The key idea common to all these designs will be to solve the brightness deterioration issue coming from the poor gun optics either by directly modifying it or by combining many sources with complementary properties in one unique gun system.

In the first scenario, two optical configurations will be studied. The first one will use a magnetic lens very close to the tip enabling a strong reduction of the aperture aberration. In collaboration with another Toulouse CNRS Laboratory, specialist in specific coil development (LNCMI: Laboratoire National des Champs Magnétique Intenses), the optical and mechanical design steps have been almost completed, and we will soon start manufacturing the new system. This first gun optical configuration will be described in Section 3.3. A second configuration using a self-centered quadrupole-octopole electrostatic gun lens will be also studied and described in Section 3.4. Because the optical design of this device is still in its early phases, the homemade optimization code will be necessary before any experimental work can begin.

Finally, for the second scenario one gun system will be proposed. This configuration is inspired by the illumination optics used in dynamic Secondary Ion Mass Spectrometer (SIMS) such as that proposed by the Cameca company (Clegg, 1991). Two distinct, but complementary, primary ion sources are used in these instruments (for instance, cesium and duoplasmatron ion sources). These two sources are simply installed before a magnetic sector, called the primary beam magnetic field (PBMF), which is used to select which one will be the source for the illumination. In the final Section 3.5 of this project, we will briefly study the feasibility of this innovative configuration for EM instruments, for instance, combining a coherent and bright electron source, such as standard CFEG, with an intense but less bright, such as a thermionic source, on the same final instrument. This conception will require to include systems with curved optic axis in the optimization code, taking into account odd order contributions of the Hamilton's point characteristic function in the perturbation term of the eikonal method.

3.3. Development of an aberration-optimized magnetic field superimposed cold field emission gun (MCFEG)

3.3.1 Introduction

As described in Section 2.1, electrostatic gun lenses were used in almost all CFEG illumination systems. The most widely used and the simplest of all triode configurations remains the Butler configuration used in the Hitachi HF CFEG (see Section 2.4). More elaborate electrostatic system were proposed such as the pentode configuration described by Veneklasen and Siegel (1972). Tetrode configurations were also studied and implemented during the high voltage STEM project (called the MEBAHT project) in the "Laboratoire d'optique électronique du CNRS" in Toulouse, the former name of the CEMES laboratory (Hawkes, 2009). Without getting into the details of a comparison between the advantages and disadvantages of different types of electrostatic CFEG, they all have the same drawbacks. On one hand they give comparable brightness and beam current and on the other hand the beam characteristics are limited by aperture C_s and chromatic C_c aberrations of the electrostatic lens (of the order of a few *cm*). Troyon (1980) developed a 100 kV field emission gun using a pre-accelerator magnetic lens. This first original design is reported in Fig. 3.4A. Besides its high

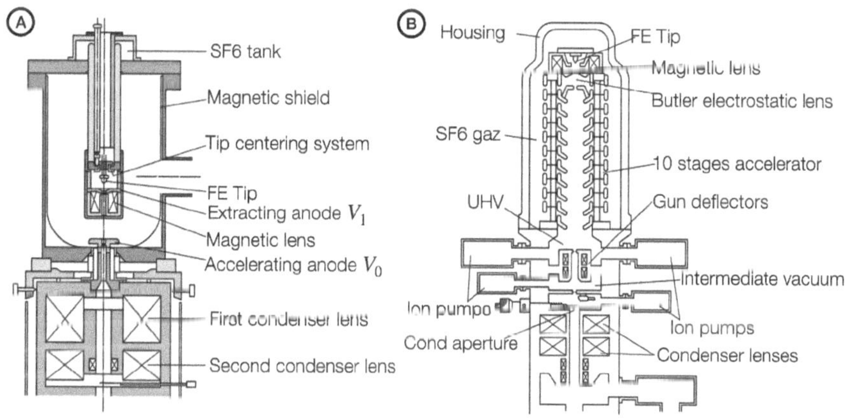

Fig. 3.4 (A) Schematic cross-section of the magnetic CFEG developed by Troyon et al. (B) Cross-section of the first Hitachi 350 kV MCFEG located before the multistage accelerator. *Panel (A): Extracted from Troyon, M., 1980. High current efficiency field emission gun system incorporating a preaccelerator magnetic lens. Its use in CTEM. Optik 57, 401. Panel (B): Extracted from Kawasaki, T., Matsuda, T., Endo, J., Tonomura, A., 1990. Observation of a 0.055 nm spacing lattice image in gold using a field emission electron microscope. Jpn. J. Appl. Phys. 29 (3A), L508.*

brightness, which is specific to all types of FEG, it provided a higher emission current density than a common electrostatic CFEG (*J* about 7 times larger at maximum brightness). In particular, it allowed a 25 times larger beam current than an electrostatic CFEG due to its low aberration coefficients in the range of $C_s \approx 2$ mm and $C_c \approx 1$ mm. They have even demonstrated that beam current close to thermionic source can be obtained without sacrificing any brightness. Moreover, they showed that the pre-accelerated magnetic lens CFEG (MCFEG) allowed the source position to be practically constant in a large range of acceleration voltage, enabling very appreciable flexibility compare to electrostatic system.

MCFEG has been incorporated in a multistage acceleration source by the Tonomura's group in the Hitachi central research laboratory (HCRL). They first installed the MCFEG in a 350 kV TEM as shown in Fig. 3.4B and on a 1 MV holography TEM (Kawazaki, 2010; Kawasaki et al., 1990, 2000). In particular, they succeeded in reaching an unrivaled probe brightness of 1.8×10^{10}A.cm^{-2}.Sr^{-1} for the 1 MV microscope. Subsequently, they developed a new holography TEM working at an acceleration voltage of 1.2 MV and fitted with a C-COR spherical aberration corrector. To align the corrector they had to stabilize the emission current of the source without any adjustment of the extraction voltage, which was not possible using the first 1 MV MCFEG configuration. Indeed using this first design, it was difficult to achieve a high vacuum around the emitter, generating current instabilities. This was mainly due to the mechanical position of the pre-accelerator magnetic lens in this first design, which restricts the vacuum conductance. This is why, for the new 1.2 MV microscope, they decided to design a new MCFEG working at an extremely high vacuum in the 10^{-10} Pa range thanks to the use of differential pumping in combination with three NEG pumps located in the magnetic lens area (Kasuya et al., 2014). The current variation using this new source over a 8 h experiment was only 5.2%. Finally, the aperture and chromatic aberration coefficients at the object plane of this new MCFEG design were calculated to be 10 and 6 mm, respectively. In both designs, the aberration coefficients are an order of magnitude smaller in the MCFEG configuration than with the original electrostatic one enabling a strong increase in beam current owing to the enlarged acceptance angle of the optics.

The multistage electrode MCFEG configuration is shown in Fig. 3.5. As in Fig. 2.3, where the evolution of the source size image (real or virtual) was described for an electrostatic CFEG with multistage acceleration, we have shown the same evolution with respect to aberrations, magnification, and diffraction for a MCFEG assembly consisting of an electrostatic lens, a multistage accelerator and one condenser lens. To simplify the mathematical

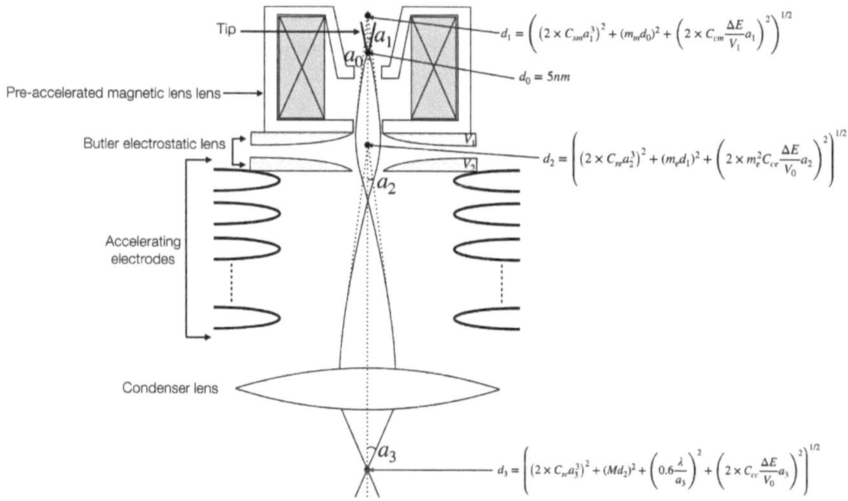

Fig. 3.5 Outline of a multistage acceleration CFEG equipped with a pre-accelerating magnetic lens showing the evolution of the source image size from the virtual source size in the emission area to the formation of the focused electron probe in the sample plane. m_m, m_e, and M are, respectively, the linear magnifications of the pre-accelerated magnetic lens, the electrostatic lens and the condenser lens. C_{sm}, C_{cm} the spherical and chromatic aberrations coefficients of the pre-accelerated magnetic lens. C_{se}, C_{ce} the spherical and chromatic aberrations coefficients of the electrostatic lens. C_{sc}, C_{cc} the spherical and chromatic aberrations coefficients of the condenser lens.

expressions all terms arising from the Lagrange–Helmoltz paraxial invariant used to express angular magnification, as in Eq. (2.7), are also hidden in the aberration coefficients. The Butler lens in Fig. 2.3 is simply replaced by the magnetic lens in Fig. 3.5. The latter is also used to extract the electrons from the tip. Indeed, in both original designs of Troyon and HHT, the upper pole piece plays the role of the first extracting anode held at the potential V_1, the axial magnetic field will then focus the electrons in a first cross over which can be real as in Troyon's design or virtual in the HHT design (Kawasaki et al., 1990; Troyon, 1980). The lower pole piece of the magnetic lens and the first accelerating anode held at a potential V_2 constitute the new Butler lens. The excitation of this Butler lens is, however, weaker than the original electrostatic configuration. The combined action of the magnetic lens and this remaining weak Butler lens will focus the image of the source at the object nodal point of the accelerator which should be located very close to the center of the Butler lens. The accelerator will then accelerate the beam at constant angular magnification so that the image emerges with the same angular divergence a_2 from the image nodal point of the accelerator located in the d_2 virtual cross-over in Fig. 2.3. This optical configuration offers two main advantages. First, the weak electrostatic lens aberrations are completely negligible compared to the magnetic lens ones, mainly due to higher magnification of the magnetic lens. Moreover, the positions of both nodal points remain relatively constant over a large range of voltage ratios V_0/V_1 between acceleration and extraction. Changing V_0 will then require only a slight adjustment of the magnetic lens excitation without any change in the Butler lens excitation.

The evolution of the probe size from the virtual source d_0 located in the tip area to the Gaussian probe size d_3 has been estimated by Kawasaki et al. using the same mathematical expressions given in Fig. 3.5 (Kawasaki et al., 2019). In particular, they derived very useful relations between the probe size d and the probe current I_p under the optimum beam angle condition that gives the smallest probe size for a given beam current I_p. They studied the dependence of this relationship on the linear magnification of the MCFEG m_m, the Butler lens m_e and the condenser lenses M. Depending on the probe current I_p, the evolution of these magnifications terms will balance the contribution of each aberrations term.

Thus, probe size d is proportional to $C_{sc}^{1/4} \times I_p^{3/8}$ for probe currents in the range $0.3\,\text{nA} < I_p < 3\,\text{nA}$. In this range d will then be limited by the condenser lens aperture aberration C_{sc} and not by the gun ones.

The probe size d is proportional to $C_{sc}^{1/4}\left(C_{cm} + \frac{C_{ce}}{m_m^2}\right)^{3/4} \times I_p^{3/4}$ for a probe current range $3\,\text{nA} < I_p$. In this case, the chromatic aberrations of the MCFEG C_{cm} and the Butler lens C_{ce} become dominant.

When I_p becomes much larger than 3 nA, the spherical aberrations of the MCFEG C_{sm} and the Butler lens C_{se} become dominant and d becomes proportional to $C_{sc}^{1/4}\left(C_{sm} + \frac{C_{se}}{m_m^4}\right)^{3/4} \times I_p^{3/2}$. This corresponds to the same situation as the conventional CFEG case without a magnetic gun lens and is known as the "angular current density limited region." In practice, this region is never reached as the probe current required for most experiments is usually located close or below 3 nA.

Thus, unlike the electrostatic CFEG, a multistage accelerator source equipped with a MCFEG assembly will never be limited by the geometrical aberrations of the gun (Kawazaki, 2010; Kawasaki et al., 1990, 2019). This is the reason why I have selected this first solution as a promising candidate for the optics of our ultrafast CFEG thereby improving the ultrafast probe current for a given brightness. However, the original Hitachi design is impracticable in our situation owing to the presence of the laser focusing optics which prevents the use of the same magnetic lens configuration (see Fig. 3.4B). In collaboration with Pierre Abeilhou in CEMES and Jérôme Béard in LNCMI, we have modified the magnetic circuit making it possible to insert the ultrafast gun assembly described in Section 2.5, while preserving the same optical working modes as previously described.

3.3.2 Ultrafast MCFEG source: First results and next steps

The first version of the Ultrafast-MCFEG (UMCFEG) mechanical configuration is shown in Fig. 3.6 and was adapted to the HF2000 CFEG described in detail in Section 2.4. The final UMCFEG assembly is installed inside a DN-CF100 chamber attached to the DN-CF63 vacuum port located at the top of the HF2000 accelerator. Various DN-CF40 vacuum flanges have been added to allow us to install NEG pumps around the tip to optimize the vacuum conductance as is the last MCFEG configuration described by HHT (Kawasaki et al., 2019). These ports are also used to introduce the coil excitation current thanks to standard metal-to-ceramic electric feedthroughs.

To open the space around the tip, we chose first to install the coil below the tip. We have also designed a magnetic circuit with an asymmetric pole

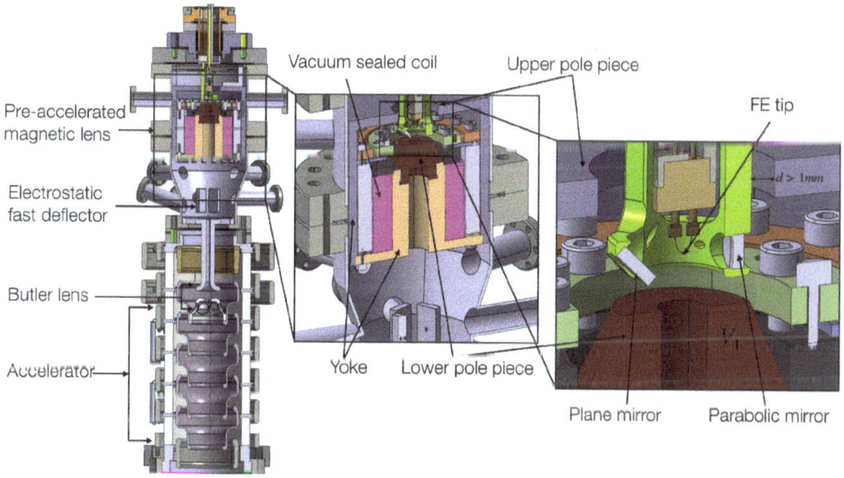

Fig. 3.6 Mechanical configuration of the new magnetic field superimposed ultrafast CFE source. A pre-accelerating magnetic lens is installed very close to the tip. The magnetic lens coil is sealed inside a dedicated chamber and is then isolated from the ultra high vacuum needed around the tip. Finally, the magnetic circuit has been designed asymmetric allowing to insert the ultrafast tip assembly (tip+mirror holder) and enabling to superimpose the tip apex with the axial magnetic field.

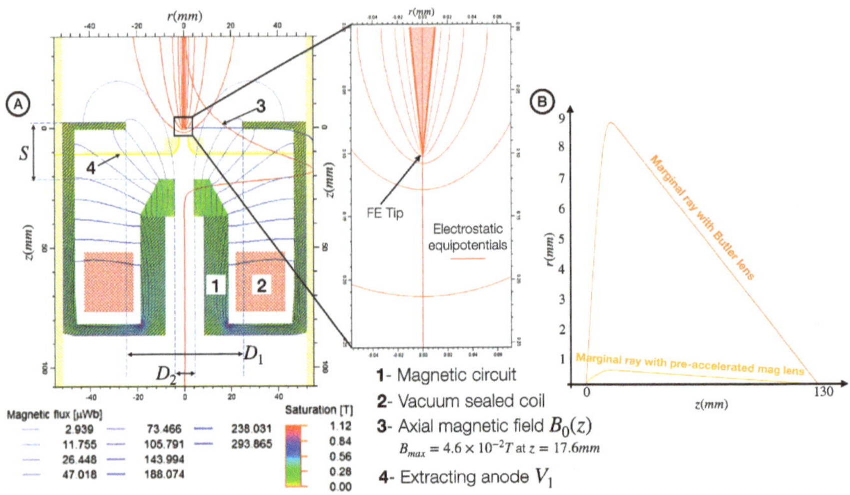

Fig. 3.7 (A) Electrostatic and magnetostatic fields configuration in the UMCFEG assembly computed using FEM thanks to EOD (Lencová and Zlámal, 2008). The configuration used to perform this simulation is $D_1 = 50$ mm, $z_t = 0$ mm, $D_2 = 8$ mm, $S = 20$ mm, $V_1 = 4$ kV and the coil excitation has been set to 1.5 A mm^{-2}. (B) Marginal paraxial rays computed using EOD inside a CFEG considering the beam focalization in a real cross-over located at $z = 130$ mm using a standard electrostatic Butler gun lens (orange) or using the new UMCFEG system (yellow).

piece configuration. Hence, the upper pole-piece hole diameter (D_1 in Fig. 3.7) should be large enough for the mirror holder to pass through while keeping the possibility of adjusting its mechanical alignment relative to the center of the lower pole-piece hole diameter (D_2). The tip + mirror holder assembly can be moved through ± 1 mm without any interference with the edge of the upper pole-piece. The tip is then located inside the pole piece gap (distance S in Fig. 3.7). As for the conventional HF2000 gun (see Fig. 2.7), the tip holder is attached to the extractor holder (DN-CF40 to DN-CF63 linked through a vacuum bellow). Hence, using the extractor holder, with which the distance between the tip and the extraction anode is selected, the tip position z_t inside the gap S of the MCFEG could be easily adjusted. The magnetic circuit (yoke + pole pieces) will be manufactured in pure iron.

Special attention has been paid to the coil design. First of all, since a water cooling system cannot be used, the coil assembly has been designed to maximize thermal surface contacts inside and outside the winding. Wire material, section size and geometry, insulating material, and general geometry of the coil have been selected by the LNCMI accustomed to complex coil design (Beard et al., 2018). Using the design presented in Fig. 3.6, the coil excitation is limited to 1.5 A mm^{-2} which will be sufficient as shown by the optical properties reported in Fig. 3.7. Such a limited excitation will be responsible for a weak increase of the temperature in the range of a few degrees Celsius. In order to prevent complex vacuum problems around the tip we choose to enclose the coil winding in a custom-built chamber isolated from the vacuum. The inner part of the coil chamber will remain at atmospheric pressure; outside the source is maintained under UHV in the range of low 10^{-9} Pa. The inner wall of this chamber will correspond to the magnetic circuit, while the other parts have been designed in non-magnetic materials to prevent perturbations of the magnetic field. The entire chamber assembly could be removed to install the coil, prepare the electric connection using the metal-to-ceramic feedthrough, and seal it before installation in the DN-CF100 general source chamber located on top of the accelerator. The tip assembly will be installed last, and the tip alignment inside the upper pole-piece hole and relative to the gap can be adjusted using the original extractor holder.

In order to achieve the appropriate vacuum level, a new baking procedure will have to be developed with a maximum baking temperature of 150 °C corresponding to the limit given by the coil insulator. A new and dedicated high-voltage conditioning procedure will be also required. Regarding the electric configuration, the gun housing and the high voltage cable will have to be modified. The metallic housing originally developed for the ultrafast electron source (see Fig. 2.16) will be ideal for this task as its modification should be straight-forward thanks to its modular construction. However, the high voltage cable, as well as the two ceramic plugs located on each side of the cable, i.e., the FE tank side and the housing side, will have to be modified. These modifications are required to bring the coil excitation current NI in the tip area in addition to the usual signals such as flash voltage V_F, extraction voltage V_1 and focusing voltage V_2 all referenced to the acceleration voltage V_0 (N is the coil number of turns and I the excitation current). Thus, two additional high-voltage referenced wires will be necessary inside the cable, and the coil power supply will have to be installed inside the FE tank as its ground will be also referenced to the V_0 voltage such as the other power supplies located inside the same tank.

Optical properties of the UMCFEG has been studied using the EOD software and first results are reported in Fig. 3.7. Field calculation in EOD is performed using the finite element method. Simulating the entire UMCFEG field distribution obliged us to combine the computation of the electrostatic and the magnetostatic potentials throughout the source. Therefore, our model generates an electrostatic mesh superimposed to the magnetostatic one and EOD provides the paraxial solutions and the primary Seidel aberrations coefficients (Lencová and Zlámal, 2008). Electrostatic (red) and magnetostatic (blue) equipotentials can be seen in the FEM simulation shown in Fig. 3.7A. The mesh size has been also adapted to the fine dimensions of the tip apex. The axial magnetic field distribution is also shown in Fig. 3.7A and is clearly asymmetric owing to the asymmetric configuration of the pole piece. A maximum value of 4.6×10^{-2} T has been obtained located close to the lower pole piece. The simulation of Fig. 3.7A has been carried using $S=20$ mm, $D_1=50$ mm and $D_2=8$ mm. The tip position $z_t=0$ mm has been chosen to coincide with the upper pole-piece bottom surface. An additional non-magnetic extractor anode is included in the magnetic lens gap (shown in yellow in Fig. 3.7A). This electrode is not present in the mechanical design shown in Fig. 3.6 as the optimum tip position relative to the magnetic lens pole piece was not yet known. First series of simulations was performed by Antoine Salih-Alj varying the tip height z_t starting from $z_t=0$ mm and going down inside the magnetic lens gap, while keeping the tip anode distance to 4 mm. Different D_1 values have been also considered in an attempt to determine the optimum diameter within a range given by a minimum value of $D_{1min}=22$ mm, corresponding to the diameter of the mirror assembly ±1 mm, and a maximum value determined by the magnetic field strength available, knowing that the excitation shouldn't be higher than 1.5 A mm^{-2}. An optimum tip position $z_t=8$ mm and a diameter $D_1=30$ mm has been found to be a satisfactory compromise. Using this configuration, a linear magnification $m_m\approx8$ has been found together with surprisingly small aperture and chromatic aberration coefficients: $C_{sm}=10\,\mu$m and $C_{cm}=5\,\mu$m. Real tracing simulations either using EOD or SIMION have also confirmed these values. Even if these numbers seems very promising compare to previous configurations developed by Troyon and HHT, my priority will be first to confirmed them using our homemade software described in Section 3.2 before starting the construction of the new UMCFEG.

The marginal ray paths estimated with either the electrostatic CFEG or the UMCFEG clearly show why our pre-accelerated magnetic lens technology improves the gun optical properties. A comparison between the two paraxial rays has been shown in Fig. 3.7B with the gun lens excited to create a real cross-over located 130 mm below the tip for the two configurations. We can clearly observe a strong decrease of the angular magnification using the UMCFEG ($m_{am} \approx 5 \times 10^{-3}$) resulting from the "confinement" effect of the magnetic lens while the electrons are emitted and accelerated. A small angular magnification leads to higher linear magnification m_m, thanks to the Lagrange–Helmholtz invariant, as well as a smaller contribution of the aperture aberration.

3.3.3 Discussion

Improving the range of available ultrafast CFEG beam current while keeping its brightness unchanged is possible by finding new gun optical configurations. Following the pioneering and encouraging work of Troyon for low-acceleration voltage TEM and HHT for higher voltages, we are designing a new source using a pre-accelerated magnetic lens. The design has been realized to respect the specific requirements of ultrafast technology (space, vacuum conductance, etc.), but the new configuration will also be used to improve the CcNT emitter optical properties. Currently, each geometrical parameter of the system (D_1, D_2, S, z_t) is being explored using EOD. Any promising aberration properties will then be confirmed using our homemade software. In the meantime, the mechanical configuration will have to be properly finalized including the additional extraction anode made of non-magnetic material. The next step will be to manufacture the coil as well as the gun DN-CF100 UHV chamber described in Fig. 3.6. A new metal housing will have to be produced taking care of the new electric feedthroughs. The new high voltage system, i.e., HV cable, ceramics, and V_0 referenced power supply, will be produced and the final source will be tested on a dedicated UHV bench originally developed to test the first ultrafast CFEG prototype. This bench is fitted with an HF2000 accelerator, two pairs of Grigson scanning coils, a Faraday cup and a camera system. Tests will be performed using the ultrafast CFEG assembly and the CcNT tips together with the UMCFEG optic. If the results are satisfactory the system will be retrofitted to the IUMi microscope to improve the amount of beam current available for future ultrafast coherent TEM experiments.

3.4. Improving an electrostatic gun lens by means of multipolar optics: From aberration optimization to aberration correction

The design described above will improve the optical properties of the electron source. Because the system will still rely on rotationally symmetric optical elements, the aberrations cannot be fully compensated. Canceling the unavoidable aperture and/or chromatic aberrations in a static CPO system without space charge requires the use of multipole elements or an electron mirror system (Hawkes and Kasper, 2018). The quadrupole–octopole solution, briefly introduced in Section 1.1.1, seems the most attractive configuration for application as a source optics corrector.

Quadrupole lenses have paraxial properties as well as aberrations described extensively in the work of Hawkes (1965, 1966). Ideally such a lens consists of four hyperbolically shaped pole faces or electrodes. Quadrupole lenses focus the beam in one plane and defocus in the perpendicular one. Thus, several lenses must normally be combined for a useful lens system. Since hyperbolic pole faces or electrodes are difficult to fabricate, the electrodes are usually approximated by cylinders such as the one shown in Fig. 3.9A for the case of an electrostatic quadrupole.

A symmetrically excited octopole has no paraxial effect upon the Gaussian imagery of a system containing round lenses or quadrupole lenses. This means that the second-order point characteristic function $S^{(2)} = \int m^{(2)} dz$ of an octopole will be zero. However, the octopole field will introduce fourth order point characteristic function $S^{(4)} = \int m^{(4)} dz$ resulting in a fourfold third-order deformation of a rotationally symmetric pencil of rays. Hence, depending on the azimuth direction inside the octopole, the sign of the aperture aberration will be different. So, contrary to a rotationally symmetric system, the aperture aberration of an octopole can be chosen opposite to that of the optical element we would like to compensate if the beam is focused along the proper azimuth of the octopole. This focusing along one direction independent from the other direction can be obtained using a quadrupole system.

Optical systems based on quadrupole elements normally consist of two or more quadrupole lenses known as *quadrupole multiplets*. For instance, a quadrupole doublet consists of two quadrupole lenses, one of which focuses in the x-direction while the other is defocusing in this direction. Tuning electrodes potentials or magnetic flux densities at the pole faces allows two conditions for the paraxial properties in the plane (xOz) and (yOz) to be satisfied. For instance, it is possible to create afocal condition in the first plane, while stigmatic condition along the other plane is created, generating a line focus in the gaussian plane along (xOz). A quadrupole doublet is, however, not very flexible, and a more commonly used quadrupole lens arrangement as a focusing element is a quadrupole triplet consisting of three lenses (Wollnik, 2014). The most attractive quadrupole multiplet is known as a *Russian quadruplet* and consists of two identical electrostatic or magnetic quadrupole doublets that are turned back-to-back to each other and rotated with respect to each other about the optic axis by 90°. Russian quadruplets systems also enable the compensation of the third-order aperture aberrations by incorporating octopole fields in specific planes located between each quadrupole. The orientation of the octopole will have to be chosen carefully to introduce the desired effect on the final aperture aberration. Four aperture aberration coefficients have to be considered in this kind of systems. Following Eq. (1.3), we have $(x|aaa)$, $(x|abb)$, $(y|baa)$, and $(y|bbb)$.

The simplest type of quadrupole–octopole (qpol/opol) corrector consists of a Russian quadruplet and three octopoles located in the mid plane between each quadrupoles (Deltrap, 1964). Solid lines in Fig. 3.8 shows one paraxial ray separated along the (xOz) and (yOz) planes after crossing the quadruplet. The quadrupole excitations are chosen to produce line foci at the octopoles O_1 and O_3. The aberration coefficient $(x|aaa)$ is corrected by exciting octopole O_1. In the same way, $(y|bbb)$ is corrected by exciting octopole O_3. Octopole O_2, located at the mid plane of the full system, is used to correct both $(x|abb)$ and $(y|baa)$. In order to correct the aperture aberration of a rotationally symmetrical lens, i.e., the spherical aberration C_s, the following condition must be fulfilled by the quadrupoles–octopoles system described in Fig. 3.8:

$$(x|aaa) = (x|abb) = (y|baa) = (y|bbb) = -C_s \tag{3.37}$$

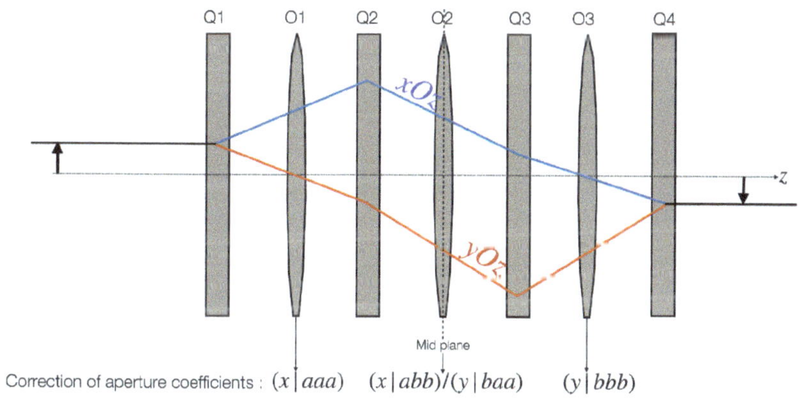

Fig. 3.8 Quadrupoles and octopoles multistage configuration used to correct primary aperture aberration of rotationally symmetrical lens (Q_1–Q_4: quadrupoles; O_1O_3: octopoles). One incoming fundamental paraxial ray has been also reported, and its propagation inside the system is shown for the two perpendicular planes (xOz) and (yOz).

Correcting the spherical aberration of a magnetic objective lens using this configuration has been realized for commercial dedicated STEM instruments by Krivanek et al. (Dellby et al., 2001). In that case, the system used a combination of magnetic quadrupoles and octopole elements. As pointed out by Deltrap already in 1960, slight misalignment in this kind of system not only results in severe degradation of correction properties but also brings in mechanical aberrations as an extra disadvantage (Deltrap, 1964). Hence, to allow for a reliable alignment of the final system this corrector consists of four-stage duodecapoles and three octopoles. Adjusting the exact alignment of the four-stage quadrupole fields is possible using the duodecapoles elements (Brown et al., 2015). Furthermore, quadrupole–octopole correction of the axial chromatic aberration is also possible using a combination of magnetic and electrostatic quadrupoles. Such a compound lens possesses first-order focusing properties of an ordinary quadrupole lens, but depending on the electrostatic and magnetic field strengths it may be achromatic or exhibit negative chromatic aberration (Yavor et al., 1964). Aperture and axial chromatic aberrations corrector systems have been also developed for TEM and SEM objective lenses following the pioneering work of the "Darmstadt project" (Rose, 1971a,b).

Compensating the aperture aberration of a CFEG Butler lens should be possible by incorporating a qpol/opol system between the extracting electrode and the focusing one located before the accelerator. However, previous configurations based on complex magnetic elements such as duodecapoles are not adapted to the pre-accelerating gun environment: too cumbersome, not easily compatible with UHV environment, too many of V_0-referenced signals will have to be injected in the gun area (specific high voltage cable, new housing and high voltage ceramics plugs, etc.) Ideally, carrying out the correction of the Butler lens should be performed using simple electrostatic qpol/opol system but the specifications required for the mechanical alignment of the elements will make the system unmanageable.

A simplified, and compact, electrostatic lens system that generates a quadrupole field and an self-aligned octopole field has been proposed by Okayama et al. The new lens unit, called self-aligned quadrupole correction lens, consists of an electrostatic quadrupole and aperture electrode as reported in Fig. 3.9A. An octopole field is automatically created within the quadrupole field by supplying a voltage to the aperture electrode (Okayama and Kawakatsu, 1982). The potential distribution of this lens, calculated by solving the Laplace's equation using FOFEM and performing its

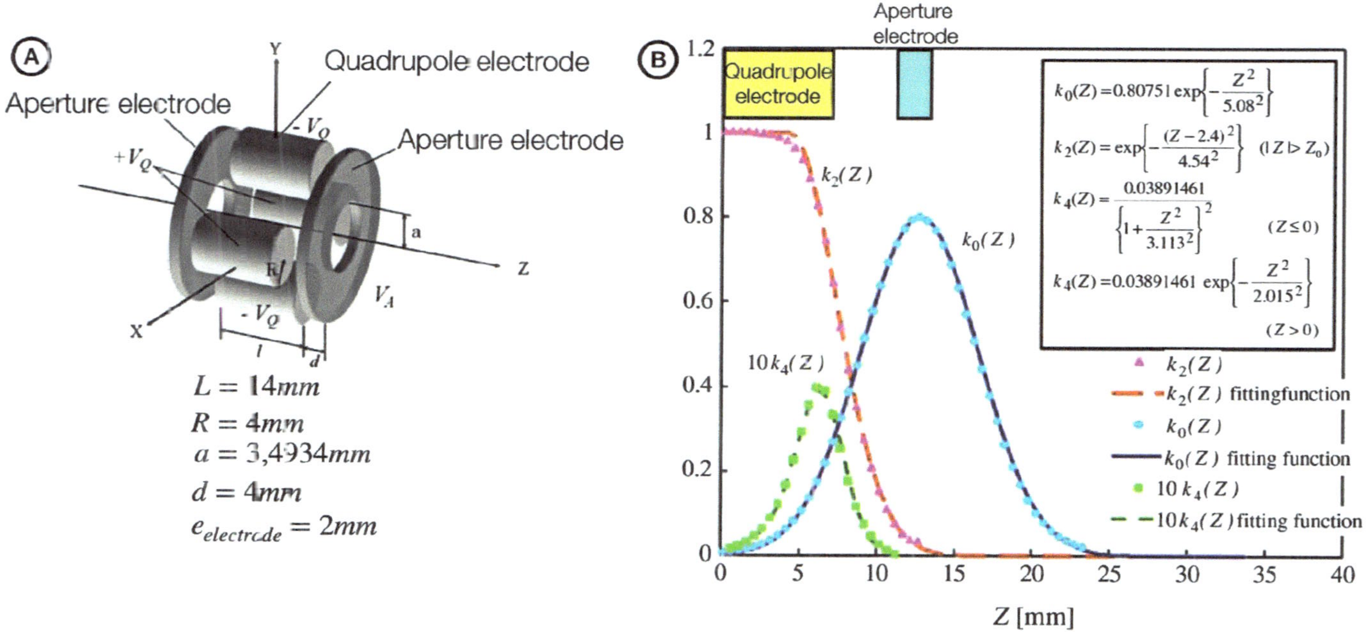

Fig. 3.9 (A) Basic arrangement of self-aligned quadrupole correction lens. $e_{electrode}$ corresponds to the thickness of the aperture electrode. (B) Characteristic functions of round-lens $k_0(z)$, quadrupole $k_2(z)$ and octopole $k_4(z)$ components of the self-aligned quadrupole correction lens. Calculations are performed using the geometrical parameters defined in panel (A). *Extracted from Tamura, K., Okayama, S., Shimizu, R., 2010. Third-order spherical aberration correction using multistage self-aligned quadrupole correction-lens systems. J. Electron Microsc. 59 (3), 197–206.*

Fourier decomposition using MEBS software, are reported in Fig. 3.9B (Chu and Munro, 1982; Tamura et al., 2010). When the aperture electrode is aligned with the electrostatic quadrupole and a voltage is supplied thereto, an octopole field $u_4(z)$ (green curve in Fig. 3.9B) is created by the fringing effect of the quadrupole field $u_2(z)$ (pink curve in Fig. 3.9B). Additionally a rotationally symmetric field $u_0(z)$ is also created centered on the aperture electrode (blue curve in Fig. 3.9B).

Each potential distribution $u_0(z)$, $u_2(z)$, and $u_4(z)$ can be expressed by using a characteristic function $k_0(z)$, $k_2(z)$, and $k_4(z)$. They correspond to the Fourier contributions of the round lens, the quadrupole lens and the octopole lens to the total electrostatic field $U(\vec{r})$ previously introduced in Eq. (3.6):

$$u_0(z) = V_A k_0(z) \tag{3.38}$$

$$u_2(z) = \frac{V_Q}{a^2} k_2(z) \tag{3.39}$$

$$u_4(z) = \frac{V_A}{a^4} k_4(z) \tag{3.40}$$

where V_A and V_Q (in volts) correspond to the aperture electrode potential and the quadrupole potential, respectively. All geometrical parameters, such as the aperture electrode radius a, are given in Fig. 3.10A.

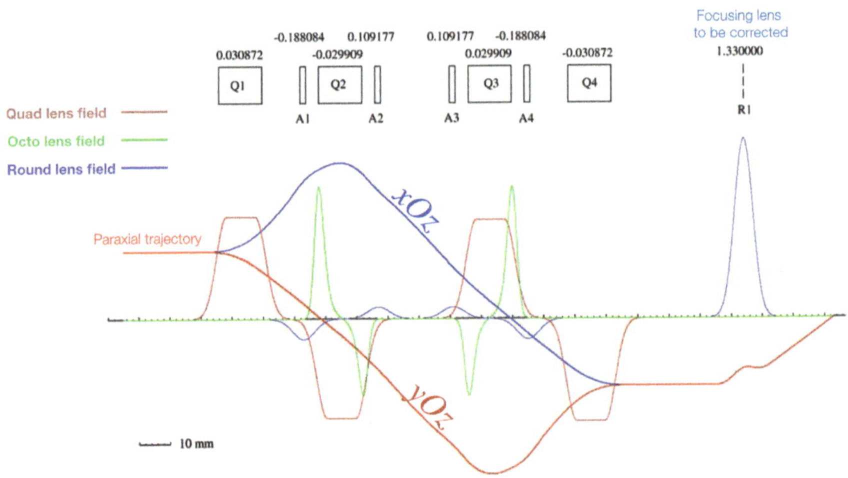

Fig. 3.10 Potentials distribution and paraxial beam trajectory in the four-stages self-aligned quadrupole correction-lens system and a round focusing lens. Q_1–Q_4 and A_1–A_4 correspond to the quadrupole and aperture electrodes, respectively. *Extracted from Tamura, K., Okayama, S., Shimizu, R., 2010. Third-order spherical aberration correction using multistage self-aligned quadrupole correction-lens systems. J. Electron Microsc. 59 (3), 197–206.*

This lens has many advantages regarding our target application. First of all this is a compact electrostatic component compatible with the gun environment (baking/UHV compatible). Furthermore, the octopole field is automatically created and aligned with the quadrupole field adjusting only one signal corresponding to the excitation of the aperture electrode. This will significantly reduce the alignment requirements of traditional qpol/opol systems and will also reduce the number of V_0 referenced signals that must be driven to the gun area. Finally, like the magnetic field in the previous MCFEG setup, the quadrupole and round-lens potentials might be employed to adjust the paraxial properties of the original Butler lens. The correction of the Butler lens aperture aberration will be possible by combining four self-aligned quadrupole correction lenses assembled to generate the same paraxial behavior as the original system reported in Fig. 3.8. Fig. 3.10 reports the simulated beam trajectory in the (xOz) and (yOz) planes together with the potential distribution of a four-stage self-aligned quadrupole correction-lens system corresponding to a standard qpol/opol corrector. In this simulation, the four-stage system is located before the focusing lens corresponding in our case to the Butler lens. This simulation has been extracted from the paper of Tamura et al. The excitations of each element (quadrupoles: Q_1, Q_2, Q_3, Q_4 and the aperture electrodes A_1, A_2, A_3, A_4) reported above them in Fig. 3.10 are calculated in order to correct the focusing lens third-order spherical aberration of $C_s = -890.4\,\text{mm}$ (Tamura et al., 2010). The total distance from the center of the first quadrupole to the fourth one is only 112 mm which is quite reasonable if we plan to install it between the extracting electrode and the focusing electrode of the HF2000 CFEG. Mechanically a dedicated interface will have to be designed and attached to the top of the accelerator using the DN-CF63 port. Nevertheless, increasing the size of the accelerator by $\approx 100\,mm$ should be compatible with the metallic housing designed originally for the ultrafast CFEG (see Section 2.5). Electrically, all the system must be referenced to the extracting anode voltage (V_1). Thus, the new Butler electrostatic lens, driven by the ratio $R = V_2/V_1$, will be generated by the voltage difference between the last aperture electrode of the system (not represented in Fig. 3.10) and the focusing electrode (V_2). The quadrupoles excitations are chosen to produce two-line foci at the octupole fields induced by the aperture electrodes A_1 and A_4 for correction of the aperture aberration coefficients $(x|aaa)$ and $(y|bbb)$. The aperture electrodes A_2 and A_3 are used to correct the aperture aberration coefficients $(x|abb)$ and $(y|baa)$. According to Tamura et al., correcting the spherical aberration of the focusing lens will require to supply the following voltages:

$$V_{A_1} = V_{A_4} = -0.18808 \times V_1 \tag{3.41}$$

$$V_{A_2} = V_{Q_3} = 0.10918 \times V_1 \tag{3.42}$$

$$V_{Q_1} = -V_{Q_4} = 0.030872 \times V_1 \tag{3.43}$$

$$V_{Q_2} = -V_{Q_3} = -0.029909 \times V_1 \tag{3.44}$$

where V_1 corresponds to the extracting anode voltage. Hence, this is another major advantage of this design regarding our target application. Indeed, because this configuration is exactly symmetric with respect to the mid-plane, it can be controlled by only six power supplies: two pairs of $\pm$DC voltage sources for the quadrupoles V_Q and two DC voltage sources for the aperture electrodes. Considering $V_1 = 4$ kV these power supplies will be in the range of $V_Q \approx 150$ V and $V_A \approx 500$ V which required cheap and easy technology. The main challenge is the manufacturing of the high voltage cable which will require 6 more V_0 referenced signals, and the two ceramics located in the FE tank and the gun housing sides.

3.5. Concepts behind a multisource optical system

I now present a longer term optical development. This new configuration should allow us to use different kind of sources on the same instrument. Originally multi-source optics have been introduced in the SIMS community (Secondary Ion Mass Spectrometry). SIMS derives compositional information from the solid under study by directing a focused ion beam at its surface. The ion beam directed at the surface is called *the primary ion beam*. Due to the interaction between the primary ion beam and the surface, emission of secondary atoms and molecules via sputtering will occur. Ions from the sputtered region are then extracted/accelerated to form the *secondary ion beam*. Carrying out the mass analysis of these secondary ions using a mass spectrometer is the principle of SIMS analysis (Van der Heide, 2014).

The history of SIMS started with the work of Georges Slodzian, and the instrument is sometimes called the Castaing–Slodzian spectrometer (Castaing and Slodzian, 1981). Following this pioneering work, SIMS instruments were improved by the Cameca company who finally developed the IMS series of instruments (IMS3-4-5-6-7F, IMS Wf, IMS 1270, and IMS 1280, see the excellent history of Cameca in de Chambost, 2011). The first instrument, the IMS 3F, was equipped with a single ion source which was a duoplasmatron. Very quickly, some improvements were made

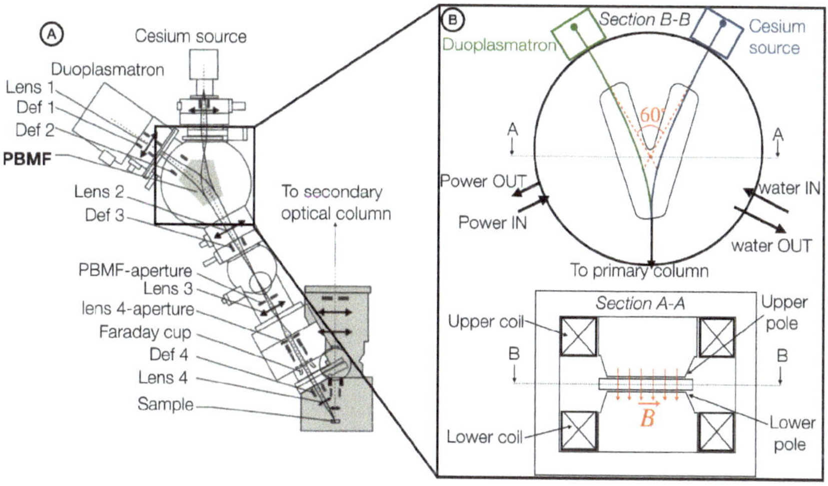

Fig. 3.11 (A) Schematic diagram of the primary column coming from a Camera IMS SIMS instrument. (B) Outline of the PBMF system fitted with two ion sources.

to this instrument. A primary beam magnet filter (PBMF) made it possible to mount two ion sources: a duoplasmatron and a cesium source. The user could select the most suitable source for the analysis. The description of this new design for the primary ion column is reported in Fig. 3.11A. Cameca also offered an electrostatic sector analyzer (ESA) instead of PBMF for specific instrument, such as the IMSWf used in the semiconductor industries. In that instrument, the cesium source is on-axis while the duoplasmatron is located at 90°. A spherical 90° ESA replaces the PBMF for switching the sources.

In Fig. 3.11B, we give a detailed description of the PBMF. The two sources are located at the entrance of a magnetic sector with an angle of 30° relative to the main optic axis of the primary ion beam column. The magnetic sector uses two coils and one large parallel pole piece generating a uniform magnetic field perpendicular to the ion beam trajectory. The ion beam will be deflected and the beam separation will depend on the mass-to-charge ratio m/Q of the primary ion beam. Modifying the coil excitation allows the user to deflect either the ion beam coming from the duoplasmatron or the one coming from the cesium source with an angle of 30° into the primary ion column optical system. The PBMF-aperture in the optical column selects the ion beam mass required for the experiments.

A first straightforward idea will be to simply transpose this design to a TEM source, replacing the duoplasmatron by a standard CFEG source and the cesium source by the ultrafast one and installing the PBMF before the accelerator. Depending on the PBMF excitation, the first or second electron beams could be injected into the accelerator. TEM alignment, or experiments requiring higher beam intensities, could then be carried out using the standard CFEG source, before switching to the ultrafast CFEG when the user needs time-resolved acquisition. However, the optical behavior of sectors (electrostatic or magnetostatic) are different from systems with a straight optic axis. Non-linear effects will have to be addressed properly to prevent a deterioration of the beam properties, which could very quickly become more dramatic than the aberrations of the Butler lens. A very good and complete description of paraxial and nonlinear properties of magnetic and electrostatic sectors can be found in the book of Wollnik (2014).

A magnetic sector is a system with curved optic axis and only one plane of symmetry. Therefore, third-order point characteristic function will have to be considered to study the nonlinear behaviors of these systems. They will induce second-order transverse aberrations such as the aperture aberration $(x|aa)$. As it is for quadrupole lens, the paraxial behavior of a magnetic sector along (xOz) or (yOz) will be also different. Additionally, linear mass and energy dispersion terms $(x|\delta)$, $(x|\gamma)$ have to be considered along the dispersion direction, chosen to be x in the following. In the paraxial approximation, assuming PBMF entrance and exit planes perpendicular to the ion beam direction (i.e., the curved z direction), the ion beam will be focused along the x direction and will not be affected along the y direction. The PBMF will then focus a round entrance beam profile into an elliptic one. Using quadrupoles, or tilted entrance and exit sector planes, this paraxial phenomenon can be compensated to retrieve a cylindrically symmetric beam (Wollnik, 2014).

In these systems, second order geometrical terms such as $(x|aa)$ and $(x|bb)$ will induce a distorsion of the paraxial beam line profile. These terms can be nevertheless compensated using one hexapole located at the entrance of the sector, for the compensation of $(x|aa)$, and a second one at the exit to compensate $(x|bb)$ term. Second-order chromatic terms such as $(x|a\delta)$, or $(x|a\gamma)$ for ion beams, are also important in curved optic axis systems, and lead to a tilt of the image plane relatively to the ion beam axis. This nonlinear chromatic effect can be also compensated using an hexapole located at the exit of the sector (Wollnik, 2014).

The paraxial behavior of an electrostatic sector will be also different along the dispersion direction x and the dispersion-free direction y. However, electrostatic sectors will separate particles according to their energy-to-charge ratio, without any mass-dependent effects so that only $(x|\delta)$ linear dispersion has to be considered. The same reasoning applies to nonlinear terms for which all chromatic terms depending on the mass difference γ are not considered. Of course, these effects are only important for ion beams. Aside from these considerations, the same paraxial and nonlinear effects have to be considered if we want to use an electrostatic sector. The second-order aperture aberrations $(x|aa)$ and $(x|bb)$ produced by electrostatic sector fields can be compensated by means of hexapole correctors, in the same manner as for magnetic sectors. However, one major advantage of electrostatic sectors is the possibility of compensating these aberrations using the fringing fields. By accelerating the beam at the entrance boundary of a sector (or by decelerating particles at the exit boundary) the fringing field decreases the second-order geometric aberrations of the sector field in the dispersion plane, and at some level of acceleration at the entrance (or deceleration at the exit), these aberrations can be compensated (Belov, 2006).

Considering our target application, the use of a pre-accelerated electrostatic system seems more suitable. The PBMF system would be more difficult to adapt in the CFEG UHV environment. Furthermore, correcting second-order effects requires hexapoles and more space will then be needed before and after the PBMF to accommodate them. Additionally, too many V_0 referenced signals would be required to excite these elements. On the contrary, the use of an electrostatic sector will be compatible with the UHV requirements. It can be designed to be more compact and finally by simply biasing the aperture electrodes properly it should be possible to minimize, or even correct, the major geometrical aberrations. The use of hexapoles located in the dispersion plane will be nevertheless needed to deal with the axial chromatic aberration $(x|a\delta)$.

Standard cylindrical or even hemispherical electrostatic sectors are usually employed to change the propagation direction of a beam but they are not easy to combine with our application (Wollnik, 2014). On the other hand, Electrostatic Quadrupole Beam Deflector (EQBD) described in Fig. 3.12A, appears to be an excellent system (Farley, 1985). Indeed, by deflecting the beam through 90°, it is then possible to install very easily three kind of sources located on the three entrance planes of an EQBD while the fourth plane will then correspond to the exit beam direction.

The EQBD consists of four hyperbolic main bending poles called quadrupole electrodes, several steering electrodes and two aperture electrodes on each side of the square housing (see Fig. 3.12A).

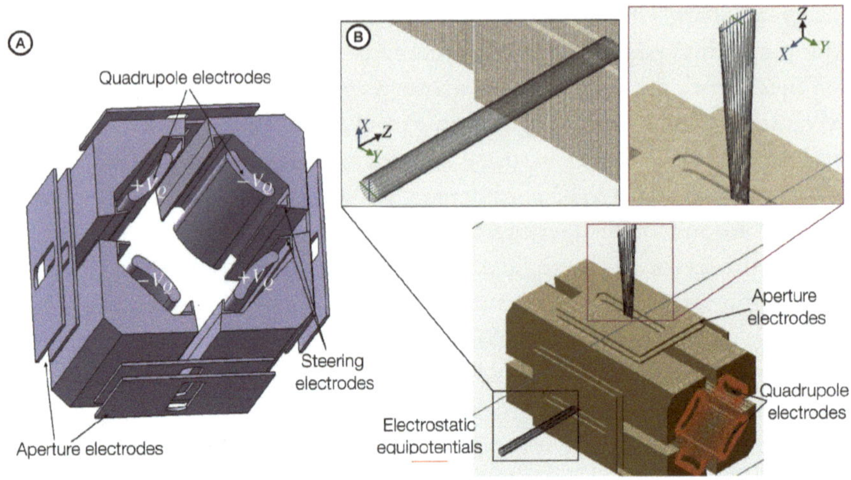

Fig. 3.12 (A) CAD design of an electrostatic quadrupole beam deflector (EQBD) seen from the edge (almost perpendicular to the beam direction). (B) SIMION simulation of 4 keV electrons beam deflected at 90° by the EQBD. Different paraxial behaviors are clearly observed comparing the beam profile along *x* and *y* at the exit (*purple square*).

Starting from the design of Fig. 3.12A, I have computed using SIMION the trajectories of 20 electrons defined by 4 keV of energy. The EQBD will have to be inserted after the V_1 electrode, i.e., above the accelerator. As reported in Fig. 3.12B, deflecting the beam through 90° requires us to apply $\pm V_Q \approx 6\,\text{kV}$ to the four quadrupole electrodes, maintaining the other electrodes at ground potential. As with standard magnetic or electrostatic sectors, due to the symmetry of the EQBD, a circular beam profile at the entrance will emerge with an elliptical beam profile. Paraxial focusing is observed only along the x direction. Quadrupole multiplets described in Section 3.4 will have to be present after the EQBD and before the accelerator. These multiplets will be used to retrieve the rotational symmetry of the beam and can act as the new electrostatic gun lens. Finally, as described in Section 3.4, if a Russian quadruplet is inserted after the EQBD and before the standard Butler lens the third-order aperture aberration of the gun can also be compensated. Second-order aberrations induced by the EQBD will have to be canceled using accelerating/decelerating fringing fields with the aid of the aperture electrodes and one additional hexapole located at the exit of the EQBD.

This system will be more cumbersome than previous MCFEG or Qpol/Opol designs. Testing it in our HF2000 bench will require the design a of new metallic gun housing. Like the previous systems, it will require a new high voltage cable, high voltage ceramics, and additional power supplies inside the FE tank. But eventually, this will make it possible to use three sources with complementary specifications (brightness, intensity, ultrafast) on the same instrument as schematically described in Fig. 3.13.

This last design is substantially more difficult than the other two owing to the incorporation of a curved-axis optical system, but it is worth the effort given the exciting possibilities offered by multi-source optics.

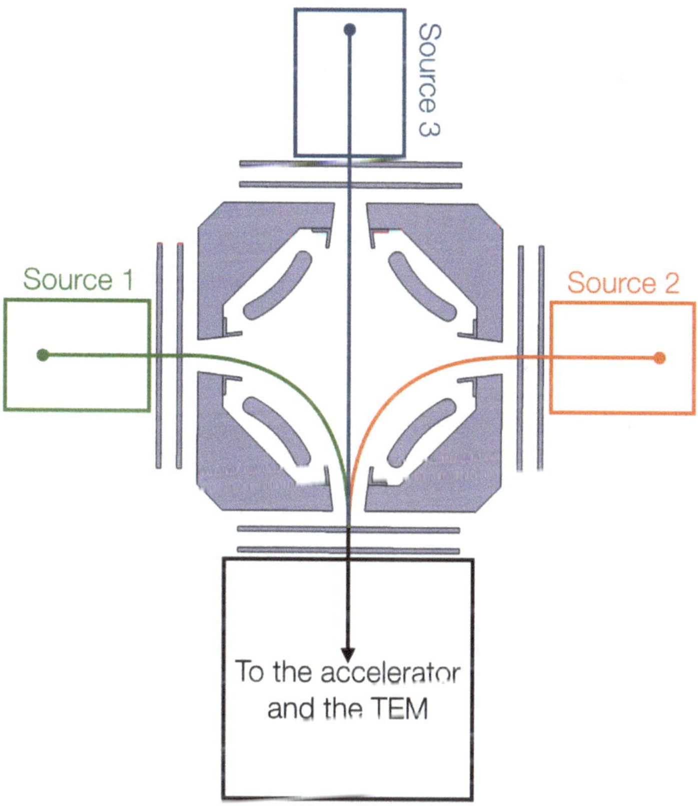

Fig. 3.13 Outline of a multi-source TEM fitted with three complementary electrons sources, an EQBD, and the TEM optics. Quadrupole multiplet and hexapole will have to be inserted after the EQBD (see text). They are not shown in this figure.

3.6. Discussion and conclusion

On ne fait jamais attention à ce qui a été fait;
on ne voit que ce qui reste à faire.

Maria Sklodowska-Curie

The development of cold field emission by Albert Crewe, professor at the university of Chicago, in the 1960s led to a profound revolution in electron microscopy. These high brightness electron sources have enabled the development of advanced probe-forming instruments, for instance, high resolution STEM such as the pioneering VG STEM, as well as highly coherent TEM allowing the sample to be coherently illuminated over a large field of view. Hitachi's pioneering work using such instruments under the leadership of Akira Tonomura has led to the development of quantitative field mapping techniques in materials at the nanoscale by means of off-axis electron holography. The advent of this kind of coherent instrument on the market made it possible to use a coherent electron beam to apply these quantitative methods to material science problems, which were first initiated at CEMES in the late 1990s and early 2000s under the impetus of Etienne Snoeck in the field of magnetism and Martin Hytch for application in stress engineering. Together with Hitachi, and thanks to our experience using the SACTEM, we were able to develop a modern instrument, the I2TEM, dedicated to the development of coherent studies. Currently, the optimization of these techniques involves more automation, more computer control of the TEM tool or new detectors, such as direct electron detectors, in order to increase the signal-to-noise ratio of the measurements carried out for instance by electron holography. Measuring ever smaller signals (magnetic, stress, etc.) while maintaining the ultimate spatial resolution offered by TEM is the goal of all these developments.

This story illustrates how a well thought-out instrumental development can give rise to a myriad of new research activities in a community like TEM. And this story was repeated with the advent of spherical aberration correctors. The history of aberration correction in electron microscopy is intimately linked to the development of electron optical theory during the 20th century. It is a very beautiful story, full of twists and turns which illustrates the evolution of a scientific adventure in the field of advanced instrumentation. From the middle of the century until the late 1990s, researchers were constantly trying to find a way to eliminate aberrations in round lenses by following theoretical ideas originally developed by optical pioneers such as Otto Scherzer, Walter Glaser, Harald Rose, Peter Hawkes, Erwin Kasper, Gertrude Rempfer, Albert Crewe, and many others. This quest was, for the large majority of these pioneers, unsuccessful and a source of great frustration in the community, so that complex developments in charged particle optics were eventually considered too risky and unfortunately abandoned in many laboratories. One of the reasons for this aversion is the international evaluation system of research which is based on the production in terms of prestigious publications, which is often incompatible with heavy instrumental developments.

Thanks to tremendous advances in computing science and electronics, aberration correction was finally achieved in TEM and STEM at the turn of the 21st century. Today, they are at the origin of the most dynamic and innovative companies in the field, CEOS and NION. Once again, the advances made in materials science but also in biology, thanks to these time-consuming instrumental developments, have been considerable. They have also led to a renewed interest in the community for the multitude of "out-of-the-box" ideas offered by the theory of electron optics.

My project aims to combine these two stories by providing innovative optical solutions to improve the properties of coherent electron sources. First, I believe this project will allow us to exploit the full potential offered by the ultra-bright CcNT and CFEG ultra-fast sources, described in the second chapter of this document. Coherent UTEM applications are currently extremely limited by the low probe current generated in the ultrafast CFEG. But I also think that these developments could be very interesting for standard FE-TEM systems. For example, it should make it possible to improve the brightness of a standard CFEG source, thus enabling it to work at lower emission current while maintaining sufficient probe current, or to couple such sources more effectively with post-acceleration monochromatic systems. The development of an optimized gun lens, coupled with shorter accelerators, should also make it possible to reduce the number of condensers, while maintaining an optimal probe size. These developments are in line with the attempts to miniaturize TEM/STEM systems without compromising spatial resolution. I am strongly convinced that correcting the source aberrations at the very beginning of the electron's "life" throughout the instrument, should maximize the benefits offered by a bright source. Indeed, when one considers that the virtual source size of a CFEG is only 5 nm, it is clear that an aplanatic gun optics with a linear magnification of only 0.01 would be sufficient to achieve a spot size of 0.5 Angstrom! This level of resolution is currently achieved by very bulky instruments equipped with several condensers, a corrector, and an objective in the case of STEM. This project is therefore part of a more general framework aimed to imagine the electron microscope of the future. In this context, I am certain that the design of the source will play a major role, together with the development of new and increasingly sensitive detectors and a privileged place for the computer science (fast simulations, fast analysis using deep learning, beam control, big data, etc.).

References

Adhikari, A., Eliason, J.K., Sun, J., Bose, R., Flannigan, D.J., Mohammed, O.F., 2017. Four-dimensional ultrafast electron microscopy: insights into an emerging technique. ACS Appl. Mater. Interfaces 9 (1), 3–16.

Aharonov, Y., Bohm, D., 1959. Significance of electromagnetic potentials in the quantum theory. Phys. Rev. 115 (3), 485–491.

Allouche, H., Monthioux, M., 2005. Chemical vapor deposition of pyrolytic carbon on carbon nanotubes. Part 2. Texture and structure. Carbon 43 (6), 1265–1278.

Allouche, H., Monthioux, M., Jacobsen, R.L., 2003. Chemical vapor deposition of pyrolytic carbon on carbon nanotubes. Carbon 41 (15), 2897–2912.

Ansys Company. Products made with Zemax define our world. From satellites orbiting the earth, to the smartphone in your pocket, the products designed by Zemax customers shape our daily interactions. https://www.zemax.com.

Arbouet, A., Houdellier, F., Marty, R., Girard, C., 2012. Interaction of an ultrashort optical pulse with a metallic nanotip: a Green dyadic approach. J. Appl. Phys. 112 (5), 053103.

Arbouet, A., Caruso, G.M., Houdellier, F., 2018. Ultrafast transmission electron microscopy: historical development, instrumentation, and applications. Adv. Imaging Electron Phys. 207, 1–72.

Barth, J.E., Kruit, P., 1996. Addition of different contributions to the charge particle probe size. Optik 101, 101.

Beard, J., Billette, J., Ferreira, N., Frings, P., Lagarrigue, J.-M., Lecouturier, F., Nicolin, J.-P., 2018. Design and tests of the 100-T triple coil at LNCMI. IEEE Trans. Appl. Supercond. 28 (3), 1–5.

Béché, A., Rouvière, J.L., Clément, L., Hartmann, J.M., 2009. Improved precision in strain measurement using nanobeam electron diffraction. Appl. Phys. Lett. 95 (12), 123114.

Belov, V.D., 2006. Immersion lenses in axially-symmetric electrostatic systems of a spherical sector deflector with second order spatial focusing. Nauchnoe Priborostroenie (Scientific Instrumentation) 16, 31.

Betcke, T., Scroggs, M., 2021. Bempp-cl: a fast Python based just-in-time compiling boundary element library. J. Open Sour. Software 6 (59), 2879.

Bird, D.M., 1989. Theory of zone axis electron diffraction. J. Electron Microsc. Tech. 13 (2), 77–97.

Bogacki, P., Shampine, L.F., 1989. A 3(2) pair of Runge-Kutta formulas. Appl. Math. Lett. 2 (4), 321–325.

Boninelli, S., Milazzo, R., Carles, R., Houdellier, F., Duffy, R., Huet, K., La Magna, A., Napolitani, E., Cristiano, F., 2018. Nanoscale measurements of phosphorous-induced lattice expansion in nanosecond laser annealed germanium. APL Mater. 6 (5), 058504.

Bormann, R., Strauch, S., Schäfer, S., Ropers, C., 2015. An ultrafast electron microscope gun driven by two-photon photoemission from a nanotip cathode. J. Appl. Phys. 118 (17), 173105.

Born, M., Wolf, E., 1999. Principles of Optics: Electromagnetic Theory of Propagation, Interference and Diffraction of Light, seventh expanded ed. Cambridge University Press, Cambridge. OCLC: 1151058062.

Bostanjoglo, O., 1989. Electron microscopy of fast processes. Adv. Electron. Electron Phys. 76, 209–279.

Bostanjoglo, O., 2002. High-speed electron microscopy. Adv. Imaging Electron Phys. 121, 1–51.

Bostanjoglo, O., Elschner, R., Mao, Z., Nink, T., Weingärtner, M., 2000. Nanosecond electron microscopes. Ultramicroscopy 81 (3–4), 141–147.
Boudarham, G., 2011. Nanooptique avec des électrons rapides: métamatériaux, formulation modale de la EMLDOS pour des systèmes plasmoniques. PhD thesis, Université Pierre et Marie Curie-Paris VI, Paris.
Boureau, V., McLeod, R., Mayall, B., Cooper, D., 2018. Off-axis electron holography combining summation of hologram series with double-exposure phase-shifting: theory and application. Ultramicroscopy 193, 52–63.
Boureau, V., Durand, A., Gergaud, P., Le Cunff, D., Wormington, M., Rouchon, D., Claverie, A., Benoit, D., Hÿtch, M., 2020. Dark-field electron holography as a recording of crystal diffraction in real space: a comparative study with high-resolution X-ray diffraction for strain analysis of MOSFETs. J. Appl. Crystallogr. 53 (4), 885–895.
Brown, L.M., Batson, P.E., Dellby, N., Krivanek, O.L., 2015. Brief history of the Cambridge STEM aberration correction project and its progeny. Ultramicroscopy 157, 88–90. https://doi.org/10.1016/j.ultramic.2015.06.006.
Buffa, A., Hiptmair, R., 2003. Galerkin boundary element methods for electromagnetic scattering. In: Barth, T.J., Griebel, M., Keyes, D.E., Nieminen, R.M., Roose, D., Schlick, T., Ainsworth, M., Davies, P., Duncan, D., Rynne, B., Martin, P. (Eds.), Topics in Computational Wave Propagation. Lecture Notes in Computational Science and Engineering, vol. 31. Springer Berlin Heidelberg, Berlin, Heidelberg, pp. 83–124.
Busch, H., 1926. Berechnung der Bahn von Kathodenstrahlen im axialsymmetrischen elektromagnetischen Felde. Ann. Phys. (Leipzig) 386 (25), 974–993.
Butler, J.W., 1966. Digital computer techniques in electron microscopy. In: 6th International Congress for Electron Microscopy. vol. 1, pp. 191–192.
Caruso, G.M., 2019. Development of a Coherent Ultrafast Transmission Electron Microscope Based on a Laser-Driven Cold Field Emission Source. PhD thesis, Université Toulouse 3, Toulouse.
Caruso, G.M., Houdellier, F., Abeilhou, P., Arbouet, A., 2017. Development of an ultrafast electron source based on a cold-field emission gun for ultrafast coherent TEM. Appl. Phys. Lett. 111 (2), 023101.
Caruso, G.M., Houdellier, F., Weber, S., Kociak, M., Arbouet, A., 2019. High brightness ultrafast transmission electron microscope based on a laser-driven cold-field emission source: principle and applications. Adv. Phys. X 4 (1), 1660214.
Castaing, R., Slodzian, G., 1981. Analytical microscopy by secondary ion imaging techniques. J. Phys. E: Sci. Instrum. 14 (10), 1119–1127.
Cherkashin, N., Reboh, S., Hÿtch, M.J., Claverie, A., Preobrazhenskii, V.V., Putyato, M.A., Semyagin, B.R., Chaldyshev, V.V., 2013. Determination of stress, strain, and elemental distribution within In(Ga)As quantum dots embedded in GaAs using advanced transmission electron microscopy. Appl. Phys. Lett. 102 (17), 173115.
Chu, H.C., Munro, E., 1982. Numerical analysis of electron beam lithography systems. Optik 61 (3), 213–236.
Clegg, J.B., 1991. Secondary ion mass spectrometry—a practical handbook for depth profiling and bulk impurity analysis, Wiley, New York, 1989. Surf. Interface Anal. 17 (4), 221.
Clément, L., Pantel, R., Kwakman, L.F.T., Rouvière, J.L., 2004. Strain measurements by convergent-beam electron diffraction: the importance of stress relaxation in lamella preparations. Appl. Phys. Lett. 85 (4), 651–653.
Cockcroft, J.D., Walton, E.T.S., 1932. Experiments with high velocity positive ions.—(I) Further developments in the method of obtaining high velocity positive ions. Proc. R. Soc. Lond. A Math. Phys. Eng. Sci. 136 (830), 619–630.
COMSOL AB. COMSOL Multiphysics v. 5.6. COMSOL, Stockholm, Sweden, www.comsol.com.

Crewe, A.V., Eggenberger, D.N., Wall, J., Welter, L.M., 1968. Electron gun using a field emission source. Rev. Sci. Instrum. 39 (4), 576–583.

Dahl, D.A., 2000. SIMION for the personal computer in reflection. Int. J. Mass Spectrom. 200 (1–3), 3–25.

De Broglie, L., 1923. Waves and quanta. Nature 112 (2815), 540.

de Chambost, E., 2011. A history of Cameca (1954–2009). Adv. Imaging Electron Phys. 167, 1–119.

de Jonge, N., 2009. Carbon nanotube electron sources for electron microscopes. Adv. Imaging Electron Phys. 156, 203–233.

de Jonge, N., Lamy, Y., Schoots, K., Oosterkamp, T.H., 2002. High brightness electron beam from a multi-walled carbon nanotube. Nature 420 (6914), 393–395.

de Jonge, N., Lamy, Y., Kaiser, M., 2003. Controlled mounting of individual multiwalled carbon nanotubes on support tips. Nano Lett. 3 (12), 1621–1624.

de Jonge, N., Allioux, M., Oostveen, J.T., Teo, K.B.K., Milne, W.I., 2005. Optical performance of carbon-nanotube electron sources. Phys. Rev. Lett. 94 (18), 186807.

de Knoop, L., Houdellier, F., Gatel, C., Masseboeuf, A., Monthioux, M., Hÿtch, M., 2014. Determining the work function of a carbon-cone cold-field emitter by in situ electron holography. Micron 63, 2–8.

de Knoop, L., Gatel, C., Houdellier, F., Monthioux, M., Masseboeuf, A., Snoeck, E., Hÿtch, M.J., 2015. Low-noise cold-field emission current obtained between two opposed carbon cone nanotips during *in situ* transmission electron microscope biasing. Appl. Phys. Lett. 106 (26), 263101.

Dellby, N., Krivanek, O.L., Nellist, P.D., Batson, P.E., Lupini, A.R., 2001. Progress in aberration-corrected scanning transmission electron microscopy. Microscopy 50 (3), 177–185.

Dellby, N., Bacon, N.J., Hrncirik, P., Murfitt, M.F., Skone, G.S., Szilagyi, Z.S., Krivanek, O.L., 2011. Dedicated STEM for 200 to 40 keV operation. Eur. Phys. J. Appl. Phys. 54 (3), 33505.

Deltrap, J.M.H., 1964. Correction of Spherical Aberration of Electron Lenses. PhD thesis, Cambridge University, Cambridge.

Denneulin, T., Gatel, C., Houdellier, F., Hÿtch, M.J., 2014. Projector lens and CCD camera distortions in a Hitachi HF-3300 TEM. In: Proc. 18th Int. Microsc. Conf, Prague, 2835. https://www.microscopy.cz/html/2835.html.

Denneulin, T., Wollschläger, N., Everhardt, A.S., Farokhipoor, S., Noheda, B., Snoeck, E., Hÿtch, M., 2018. Local deformation gradients in epitaxial $Pb(Zr_{0.2}Ti_{0.8})O_3$ layers investigated by transmission electron microscopy. J. Phys. Condens. Matter 30 (21), 215701.

de Ruijter, W.J., Weiss, J.K., 1993. Detection limits in quantitative off-axis electron holography. Ultramicroscopy 50 (3), 269–283.

De Wolf, I., Senez, V., Balboni, R., Armigliato, A., Frabboni, S., Cedola, A., Lagomarsino, S., 2003. Techniques for mechanical strain analysis in sub-micrometer structures: TEM/CBED, micro-Raman spectroscopy, X-ray micro-diffraction and modeling. Microelectron. Eng. 70 (2–4), 425–435.

Dupuy, J., 2021. Contrôle dynamique et optimisation des observations en microscopie électronique en transmission. PhD thesis, Université Toulouse 3, Toulouse.

Dushman, S., 1930. Thermionic emission. Rev. Mod. Phys. 2 (4), 381–476.

Egoavil, R., Gauquelin, N., Martinez, G.T., Van Aert, S., Van Tendeloo, G., Verbeeck, J., 2014. Atomic resolution mapping of phonon excitations in STEM-EELS experiments. Ultramicroscopy 147, 1–7.

Ehrenberg, W., Siday, R.E., 1949. The refractive index in electron optics and the principles of dynamics. Proc. Phys. Soc. B 62 (1), 8–21.

Einsle, J.F., Gatel, C., Masseboeuf, A., Cours, R., Bashir, M.A., Gubbins, M., Bowman, R.M., Snoeck, E., 2015. In situ electron holography of the dynamic magnetic field emanating from a hard-disk drive writer. Nano Res. 8 (4), 1241–1249.

Farley, J.W., 1985. Simple electrostatic quadrupole ion beam deflector. Rev. Sci. Instrum. 56 (9), 1834–1835.

Feist, A., Echternkamp, K.E., Schauss, J., Yalunin, S.V., Schäfer, S., Ropers, C., 2015. Quantum coherent optical phase modulation in an ultrafast transmission electron microscope. Nature 521 (7551), 200–203.

Feist, A., Bach, N., Rubiano da Silva, N., Danz, T., Möller, M., Priebe, K.E., Domröse, T., Gatzmann, J.G., Rost, S., Schauss, J., Strauch, S., Bormann, R., Sivis, M., Schäfer, S., Ropers, C., 2017. Ultrafast transmission electron microscopy using a laser-driven field emitter: femtosecond resolution with a high coherence electron beam. Ultramicroscopy 176, 63–73.

Fowler, R.H., Nordheim, L., 1928. Electron emission in intense electric fields. Proc. R. Soc. Lond. A Math. Phys. Charac. 119 (781), 173–181.

Fransen, M.J., Overwijk, M.H.F., Kruit, P., 1999a. Brightness measurements of a ZrO/W Schottky electron emitter in a transmission electron microscope. Appl. Surf. Sci. 146 (1–4), 357–362.

Fransen, M.J., Van Rooy, T.L., Tiemeijer, P.C., Overwijk, M.H.F., Faber, J.S., Kruit, P., 1999b. On the electron-optical properties of the ZrO/W Schottky electron emitter. Adv. Imaging Electron Phys. 111, 91–166.

Fredholm, I., 1903. Sur une classe d'équations fonctionnelles. Acta Math. 27 (0), 365–390.

Gabor, D., 1948. A new microscopic principle. Nature 161 (4098), 777–778.

Gammer, C., Ophus, C., Pekin, T.C., Eckert, J., Minor, A.M., 2018. Local nanoscale strain mapping of a metallic glass during *in situ* testing. Appl. Phys. Lett. 112 (17), 171905.

García de Abajo F.J., 2010. Optical excitations in electron microscopy. Rev. Mod. Phys. 82 (1), 209–275.

García de Abajo F.J., Asenjo-Garcia, A., Kociak, M., 2010. Multiphoton absorption and emission by interaction of swift electrons with evanescent light fields. Nano Lett. 10 (5), 1859–1863.

Gatel, C., 2020. Quantitative Field Mapping at the Nanoscale by Transmission Electron Microscopy. HDR, Université Toulouse 3, Toulouse.

Gatel, C., Dupuy, J., Houdellier, F., Hÿtch, M.J., 2018. Unlimited acquisition time in electron holography by automated feedback control of transmission electron microscope. Appl. Phys. Lett. 113 (13), 133102.

Glaser, W., 1952. Grundlagen der Elektronenoptik. Springer Berlin Heidelberg. OCLC: 1026931556.

Gratias, D., Portier, R., 1983. Time-like perturbation method in high-energy electron diffraction. Acta Crystallogr. A 39 (4), 576–584.

Grivet, P., 1972. Electron Optics. Part 1, Optics, second (revised) English edition. Pergamon, Oxford, OCLC: 899003663.

Gross, H. (Ed.), 2005. Handbook of Optical Systems. Wiley-VCH, Weinheim. OCLC: ocm56963682.

Hamilton, W.R., 1833. On a general method of expressing the paths of light and of the planets by the coefficients of a characteristic function. Dublin Univ. Rev. Q. Mag. 795 (1).

Hanbücken, M., Deville, J.-P., 2001. Stress and Strain in Epitaxy: Theoretical Concepts, Measurements, and Applications: Keynote Lectures of the Third Porquerolles School on Special Topics in Surface Science, Ile de Porquerolles, France, October 1–7, 2000. Elsevier Science, Amsterdam. OCLC: 162130694.

Hanszen, K.J., 1986. Method of off-axis electron holography and investigations of the phase structure in crystals. J. Phys. D: Appl. Phys. 19 (3), 373–395.

Harada, K., Tonomura, A., Togawa, Y., Akashi, T., Matsuda, T., 2004. Double-biprism electron interferometry. Appl. Phys. Lett. 84 (17), 3229–3231.

Harada, K., Matsuda, T., Tonomura, A., Akashi, T., Togawa, Y., 2006. Triple-biprism electron interferometry. J. Appl. Phys. 99 (11), 113502.

Harscher, A., Lichte, H., 1996. Experimental study of amplitude and phase detection limits in electron holography. Ultramicroscopy 64 (1–4), 57–66.

Hashimoto, T., Sato, Y., Ueki, Y., Kamino, T., Isakozawa, S., Ikematsu, Y., Uemori, R., Morikawa, H., 1991. Nano-area analysis by an energy dispersive X-ray spectrometer mounted on a field emission transmission electron microscope. Proc. EMSA 49, 352.

Hawkes, P.W., 1965. The geometrical aberrations of general electron optical systems II. The primary (third order) aberrations of orthogonal systems, and the secondary (fifth order) aberrations of round systems. Philos. Trans. R. Soc. Lond. A Math. Phys. Sci. 257 (1086), 523–552.

Hawkes, P.W., 1966. Quadrupole Optics. Springer Tracts in Modern Physics, vol. 42 Springer, Berlin, Heidelberg.

Hawkes, P.W. (Ed.), 1982. Magnetic Electron Lenses. Springer-Verlag, Berlin. OCLC: 680001010.

Hawkes, P.W., 1990. In brief. Phys. Today 43 (7), 84.

Hawkes, P.W. (Ed.), 2009. Cold field emission and the scanning transmission electron microscope. Adv. Imaging Electron Phys. 159.

Hawkes, P.W., Kasper, E., 2018. Principles of Electron Optics. Volume 1: Basic Geometrical Optics, second ed. Elsevier, London.

Hawkes, P.W., Spence, J.C.H. (Eds.), 2007. Science of Microscopy. Springer, New York, NY.

Hirsch, P.B., Howie, A., Nicholson, R.B., Pashley, D.W., Whelan, M.J., 1977. Electron Microscopy of Thin Crystals. R. E. Krieger Pub. Co, Huntington, NY.

Hommelhoff, P., Kealhofer, C., Kasevich, M.A., 2006a. Ultrafast electron pulses from a tungsten tip triggered by low-power femtosecond laser pulses. Phys. Rev. Lett. 97 (24), 247402.

Hommelhoff, P., Sortais, Y., Aghajani-Talesh, A., Kasevich, M.A., 2006b. Field emission tip as a nanometer source of free electron femtosecond pulses. Phys. Rev. Lett. 96 (7), 077401.

Houdellier, F., 2006. Contribution au développement du CBED et de l'holographie HREM pour l'analyse des déformations de couches épitaxiées. PhD thesis, INSA, Toulouse.

Houdellier, F., Arbouet, A., 2018. HC-IUMi, https://www.cemes.fr/HC-IUMi-the-creation-of-the-joint?lang=en.

Houdellier, F., Roucau, C., Clément, L., Rouvière, J.L., Casanove, M.J., 2006. Quantitative analysis of HOLZ line splitting in CBED patterns of epitaxially strained layers. Ultramicroscopy 106 (10), 951–959.

Houdellier, F., Altibelli, A., Roucau, C., Casanove, M.J., 2008a. New approach for the dynamical simulation of CBED patterns in heavily strained specimens. Ultramicroscopy 108 (5), 426–432.

Houdellier, F., Hÿtch, M., Hüe, F., Snoeck, E., 2008b. Aberration correction with the SACTEM-Toulouse: from imaging to diffraction. Adv. Imaging Electron Phys. 153, 225–259.

Houdellier, F., Jacob, D., Casanove, M.J., Roucau, C., 2008c. Effect of sample bending on diffracted intensities observed in CBED patterns of plan view strained samples. Ultramicroscopy 108 (4), 295–301.

Houdellier, F., Masseboeuf, A., Monthioux, M., Hÿtch, M.J., 2012. New carbon cone nanotip for use in a highly coherent cold field emission electron microscope. Carbon 50 (5), 2037–2044.

Houdellier, F., de Knoop, L., Gatel, C., Masseboeuf, A., Mamishin, S., Taniguchi, Y., Delmas, M., Monthioux, M., Hÿtch, M.J., Snoeck, E., 2015a. Development of TEM and SEM high brightness electron guns using cold-field emission from a carbon nanotip. Ultramicroscopy 151, 107–115.

Houdellier, F., Röder, F., Snoeck, E., 2015b. Development of splitting convergent beam electron diffraction (SCBED). Ultramicroscopy 159, 59–66.
Houdellier, F., Caruso, G.M., Weber, S., Kociak, M., Arbouet, A., 2018. Development of a high brightness ultrafast transmission electron microscope based on a laser-driven cold field emission source. Ultramicroscopy 186, 128–138.
Houdellier, F., Caruso, G.M., Weber, S., Hÿtch, M.J., Gatel, C., Arbouet, A., 2019. Optimization of off-axis electron holography performed with femtosecond electron pulses. Ultramicroscopy 202, 26–32.
Howie, A., 2012. Akira Tonomura (1942–2012). Nature 486 (7403), 324.
Hui, R., O'Sullivan, M., 2009. Fiber Optic Measurement Techniques. Elsevier.
Hÿtch, M.J., Snoeck, E., Kilaas, R., 1998. Quantitative measurement of displacement and strain fields from HREM micrographs. Ultramicroscopy 74 (3), 131–146.
Hÿtch, M., Houdellier, F., Hüe, F., Snoeck, E., 2008a. Nanoscale holographic interferometry for strain measurements in electronic devices. Nature 453 (7198), 1086–1089.
Hÿtch, M.J., Houdellier, F., Hüe, F., Snoeck, E., 2008b. PCT/FR2008/001302. https://www.hremresearch.com/Eng/plugin/HoloDarkEng.html.
Hÿtch, M.J., Houdellier, F., Hüe, F., Snoeck, E., 2011. Dark-field electron holography for the measurement of geometric phase. Ultramicroscopy 111 (8), 1328–1337.
Ibe, J.P., Bey, P.P., Brandow, S.L., Brizzolara, R.A., Burnham, N.A., DiLella, D.P., Lee, K.P., Marrian, C.R.K., Colton, R.J., 1990. On the electrochemical etching of tips for scanning tunneling microscopy. J. Vac. Sci. Technol. A 8 (4), 3570–3575.
Ierusalimschy, R., 2003. Programming in Lua. Lua.org, Rio de Janeiro.
Iida, H., Taniguchi, Y., 2016. Global collaboration with CEMES, French National Center for Scientific Research: joint project to implement aberration corrector in high-end TEM. Hitachi Rev. 65 (7), 283.
Isakozawa, S., Kashikura, Y., Sato, Y., Takahashi, Y., Ichihashi, M., Murakoshi, H., 1989. Development of a 200 kV field emission TEM. Proc. EMSA 47, 112–113.
Jagannathan, R., Khan, S.A., 2019. Quantum mechanics of charged particle beam optics: understanding devices from electron microscopes to particle accelerators. CRC Press, Boca Raton. OCLC: 680001010.
Kabius, B., Haider, M., Uhlemann, S., Schwan, E., Urban, K., Rose, H., 2002. First application of a spherical-aberration corrected transmission electron microscope in materials science. J. Electron Microsc. 51 (Suppl. 1), S51–S58.
Kamino, T., Tomita, M., 1983. Effect of FE gun in analytical electron microscope. Hitachi Technical Data "EM" Sheet 36.
Kasper, E., 1983. On the solution of integral equations arising in electron optical field calculations. Optik 64, 157.
Kasper, E., Scherle, W., 1982. On the analytical calculation of field in electron optical devices. Optik 60, 339.
Kasuya, K., Kawasaki, T., Moriya, N., Arai, M., Furutsu, T., 2014. Magnetic field superimposed cold field emission gun under extreme-high vacuum. J. Vacuum Sci. Technol. B Nanotechnol. Microelectron.: Mater. Process. Meas. Phenomena. 32 (3), 031802.
Kawazaki, T., 2010. Development and application of field-emission electron guns with superimposed magnetic fields. PhD thesis.
Kawasaki, T., Matsuda, T., Endo, J., Tonomura, A., 1990. Observation of a 0.055 nm spacing lattice image in gold using a field emission electron microscope. Jpn. J. Appl. Phys. 29 (3A), L508.
Kawasaki, T., Matsui, I., Yoshida, T., Katsuta, T., Hayashi, S., Onai, T., Furutsu, T., Myochin, K., Numata, M., Mogaki, H., Gorai, M., Akashi, T., Kamimura, O., Matsuda, T., Osakabe, N., Tonomura, A., Kitazawa, K., 2000. Development of a 1 MV field-emission transmission electron microscope. J. Electron Microsc. 49 (6), 711–718.

Kawasaki, T., Akashi, T., Kasuya, K., Shinada, H., 2019. Beam brightness and its reduction in a 1.2-MV cold field-emission transmission electron microscope. Ultramicroscopy 202, 107–113.

Kittel, C., 2005. Introduction to Solid State Physics, eighth ed. Wiley, Hoboken, NJ.

Koch, D., 2003. SF_6 properties and use in MV and HV switchgear. Technical document Schneider Electric, 188.

Kociak, M., Zagonel, L.F., 2017. Cathodoluminescence in the scanning transmission electron microscope. Ultramicroscopy 176, 112–131.

Kociak, M., Zagonel, L.F., Tencé, M., Mazzucco, S., 2018. Flexible cathodoluminescence detection system and microscope employing such a system. Patent application 20130087706 A1.

Krüger, M., Schenk, M., Hommelhoff, P., 2011. Attosecond control of electrons emitted from a nanoscale metal tip. Nature 475 (7354), 78–81.

Kubo, Y., Gatel, C., Snoeck, E., Houdellier, F., 2017. Optimising electron microscopy experiment through electron optics simulation. Ultramicroscopy 175, 67–80.

Kutta, W., 1901. Beitrag zur näherungsweisen Integration totaler Differentialgleichungen. Teubner.

LaGrange, T., Reed, B., DeHope, W., Shuttlesworth, R., Huete, G., 2011. Movie mode dynamic transmission electron microscopy (DTEM): multiple frame movies of transient states in materials with nanosecond time resolution. Microsc. Microanal. 17 (S2), 458–459.

Lambda Research Corporation, 2021. Optics references. https://lambdaresfiles.com/wp-content/uploads/support/oslo/oslo_releases/OSLOOpticsReference.pdf.

Lambda Research Corporation. Optics Software for Layout and Optimization meets today's lens design and optical engineering requirements. https://lambdares.com/oslo/.

Leavers, V.F., 1992. Shape Detection in Computer Vision Using the Hough Transform. Springer-Verlag, London; New York.

Legros, M., Ferry, O., Houdellier, F., Jacques, A., George, A., 2008. Fatigue of single crystalline silicon: mechanical behaviour and TEM observations. Mater. Sci. Eng. A 483–484, 353–364.

Lehmann, M., 2004. Influence of the elliptical illumination on acquisition and correction of coherent aberrations in high-resolution electron holography. Ultramicroscopy 100 (1–2), 9–23.

Lencová, B., Zlámal, J., 2008. A new program for the design of electron microscopes. Phys. Proc. 1 (1), 315–324.

Lichte, H., 1991. Optimum focus for taking electron holograms. Ultramicroscopy 38 (1), 13–22.

Linck, M., 2018. Demagnification setting for B COR Hitachi, Internal Document CEOS–CEMES.

Lourenço-Martins, H., Lubk, A., Kociak, M., 2021. Bridging nano-optics and condensed matter formalisms in a unified description of inelastic scattering of relativistic electron beams. SciPost Phys. 10 (2), 031.

Mamishin, S., Kubo, Y., Cours, R., Monthioux, M., Houdellier, F., 2017. 200 keV cold field emission source using carbon cone nanotip: application to scanning transmission electron microscopy. Ultramicroscopy 182, 303–307.

Massmann, I., Uhlemann, S., Muller, H., Hartel, P., Zach, J., Haider, M., Taniguchi, Y., Hoyle, D., Herring, R., 2011. Realization of the first aplanatic transmission electron microscope. Microsc. Microanal. 17 (S2), 1270–1271.

McLean, W.C.H., 2000. Strongly Elliptic Systems and Boundary Integral Equations. Cambridge University Press, Cambridge; New York.

Meury, P., 2007. Stable Finite Element Boundary Element Galerkin Schemes for Acoustic and Electromagnetic Scattering. PhD thesis, ETH Zurich, Zurich.

Möllenstedt, G., Düker, H., 1956. Beobachtungen und Messungen an Biprisma-Interferenzen mit Elektronenwellen. Z. Phys. 145 (3), 377–397.

Mompiou, F., 2015. Mécanismes de déformation plastique et Microscopie Electronique en Transmission in-situ. HDR, Université Toulouse 3, Toulouse.

Mompiou, F., Caillard, D., 2014. Dislocations and mechanical properties of icosahedral quasicrystals. C. R. Phys. 15 (1), 82–89.

Mompiou, F., Caillard, D., Legros, M., Mughrabi, H., 2012. In situ TEM observations of reverse dislocation motion upon unloading in tensile-deformed UFG aluminium. Acta Mater. 60 (8), 3402–3414.

Monthioux, M., Allouche, H., Jacobsen, R.L., 2006. Chemical vapour deposition of pyrolytic carbon on carbon nanotubes. Carbon 44 (15), 3183–3194.

Morniroli, J.P., 2002. Large-Angle Convergent-Beam Electron Diffraction (LACBED): Applications to Crystal Defects. Monograph of the French Society of Microscopies. Société Française des Microscopies, Paris. OCLC: ocm55209388.

Müller, H., Rose, H., Uhlemann, S., 2010. Electron-optical corrector for aplanatic imaging systems. Patent USPTO 7,800,076.

Müller, H., Maßmann, I., Uhlemann, S., Hartel, P., Zach, J., Haider, M., 2011. Aplanatic imaging systems for the transmission electron microscope. Nucl. Instrum. Methods Phys. Res. A 645 (1), 20–27.

Murakoshi, H., Ichihashi, M., Isakozawa, S., Sato, Y., 1992. Field emission electron device which produces a constant beam current. Patent US5134289.

Murata, H., Ohye, T., Shimoyama, H., 2004. High accuracy calculation of magnetic field by improved 3-D boundary magnetic charge method. Nucl. Instrum. Methods Phys. Res. A 519 (1–2), 175–183.

Murata, H., Ishigami, M., Shimoyama, H., 2016. Development of a boundary magnetic charge method for computing magnetic fields in a system containing saturated magnetic materials. Nucl. Instrum. Methods Phys. Res. A 806, 360–369.

Nagaoka, K., Yamashita, T., Uchiyama, S., Yamada, M., Fujii, H., Oshima, C., 1998. Monochromatic electron emission from the macroscopic quantum state of a superconductor. Nature 396 (6711), 557–559.

Okayama, S., Kawakatsu, H., 1982. A new correction lens. J. Phys. E: Sci. Instrum. 15 (5), 580–586.

Orchowski, A., Rau, W.D., Lichte, H., 1995. Electron holography surmounts resolution limit of electron microscopy. Phys. Rev. Lett. 74 (3), 399–402.

Orloff, J. (Ed.), 2009. Handbook of Charged Particle Optics, second ed. CRC Press/Taylor & Francis, Boca Raton. OCLC: ocn213765926.

Pennycook, S.J., Nellist, P.D. (Eds.), 2011. Scanning Transmission Electron Microscopy: Imaging and Analysis. Springer, New York, NY.

Press, W.H. (Ed.), 1988. Numerical Recipes in C: The Art of Scientific Computing. Cambridge University Press, Cambridge.

Reed, B.W., LaGrange, T., Shuttlesworth, R.M., Gibson, D.J., Campbell, G.H., Browning, N.D., 2010. Solving the accelerator-condenser coupling problem in a nanosecond dynamic transmission electron microscope. Rev. Sci. Instrum. 81 (5), 053706.

Reimer, L., Kohl, E., 2008. Transmission Electron Microscopy: Physics of Image Formation. Springer Series in Optical Sciences, vol. 36 Springer, New York, NY.

Ribaya, B.P., Leung, J., Brown, P., Rahman, M., Nguyen, C.V., 2008. A study on the mechanical and electrical reliability of individual carbon nanotube field emission cathodes. Nanotechnology 19 (18), 185201.

Riley, K.F., Hobson, M.P., Bence, S.J., 2006. Mathematical Methods for Physics and Engineering, third ed. Cambridge University Press, Cambridge. OCLC: 569384420.

Röder, F., Houdellier, F., Denneulin, T., Snoeck, E., Hÿtch, M., 2016. Realization of a tilted reference wave for electron holography by means of a condenser biprism. Ultramicroscopy 161, 23–40.

Ropers, C., Solli, D.R., Schulz, C.P., Lienau, C., Elsaesser, T., 2007. Localized multiphoton emission of femtosecond electron pulses from metal nanotips. Phys. Rev. Lett. 98 (4), 043907.

Rose, H., 1971a. Abbildungseigenschaften spharisch korrigierter elektronenoptische Achromate. Optik 33, 1.

Rose, H., 1971b. Elektronenoptische Aplanate. Optik 34, 285.

Rose, H., 1987. Hamiltonian magnetic optics. Nucl. Instrum. Methods Phys. Res. A 258 (3), 374–401.

Rose, H., 1990. Outline of a spherically corrected semiaplanatic medium-voltage transmission electron microscope. Optik 85, 19–24.

Rose, H.H., 2009. Geometrical Charged-Particle Optics. Springer Series in Optical Sciences, vol. 142 Springer, Berlin. OCLC: ocn268931877.

Saito, Y., Seko, K., Kinoshita, J.-i., 2005. Dynamic behavior of carbon nanotube field emitters observed by in situ transmission electron microscopy. Diamond Relat. Mater. 14 (11–12), 1843–1847.

Schenk, M., Krüger, M., Hommelhoff, P., 2010. Strong-field above-threshold photoemission from sharp metal tips. Phys. Rev. Lett. 105 (25), 257601.

Scherzer, O., 1936. Über einige Fehler von Elektronenlinsen. Z. Phys. 101 (9–10), 593–603.

Schwertfeger, W., Kasper, E., 1974. Zur numerischen Berechnung elektromagnetischer Multipolfelder. Optik 41, 160–173.

Scroggs, M.W., Betcke, T., Burman, E., Śmigaj, W., van 't Wout, E., 2017. Software frameworks for integral equations in electromagnetic scattering based on Calderón identities. Comput. Math. Appl. 74 (11), 2897–2914.

SCT Company. Société des Céramiques Techniques. https://www.sct-ceramics.com.

Septier, A. (Ed.), 1967. Focusing of Charged Particles. Academic Press, New York and London.

Shah, J., 1999. Ultrafast Spectroscopy of Semiconductors and Semiconductor Nanostructures. Springer Series in Solid-State Sciences, vol. 115 Springer Berlin Heidelberg, Berlin, Heidelberg.

Singh, G., Bücker, R., Kassier, G., Barthelmess, M., Zheng, F., Migunov, V., Kruth, M., Dunin-Borkowski, R.E., Purcell, S.T., Dwayne Miller, R.J., 2018. Fabrication and characterization of a focused ion beam milled lanthanum hexaboride based cold field electron emitter source. Appl. Phys. Lett. 113 (9), 093101.

Smith, G.E. 1997. J.J Thomson and the electron: 1897–1899 an introduction. Chem. Educ. 2 (6), 1–42.

Snoeck, E., Müller, H., Haider, M., 2006. Using a CEOS-objective lens corrector as a pseudo Lorentz lens in a Tecnai F20 TEM. In: Proc. 16th International Microscopy Congress, 730.

Spence, J.C.H., 2013. High-Resolution Electron Microscopy. Oxford University Press.

Spence, J.C.H., Zuo, J.M., 2013. Electron Microdiffraction. Springer Science + Business Media, LLC, New York, NY. OCLC: 921057646.

Stadelmann, P.A., 1987. EMS—a software package for electron diffraction analysis and HREM image simulation in materials science. Ultramicroscopy 21 (2), 131–145.

Steinbach, O., 2008. Numerical Approximation Methods for Elliptic Boundary Value Problems. Springer, New York, NY.

Sturrock, P.A., 1951. Perturbation characteristic functions and their application to electron optics. Proc. R. Soc. Lond. A 210 (1101), 269–289.

Swanson, L., Schwind, G., 2008. Review of ZrO/W Schottky cathode. In: Handbook of Charged Particle Optics, second ed. CRC Press, pp. 1–28.

Swanson, L.W., Schwind, G.A., 2009. Chapter 2: A review of the cold-field electron cathode. Adv. Imaging Electron Phys. 159, 63–100.

Szilagyi, M., 1988. Electron and Ion Optics. Springer US, Boston, MA.

Tamura, K., Okayama, S., Shimizu, R., 2010. Third-order spherical aberration correction using multistage self-aligned quadrupole correction-lens systems. J. Electron Microsc. 59 (3), 197–206.

Tanigaki, T., Inada, Y., Aizawa, S., Suzuki, T., Park, H.S., Matsuda, T., Taniyama, A., Shindo, D., Tonomura, A., 2012. Split-illumination electron holography. Appl. Phys. Lett. 101 (4), 043101.

Tanigaki, T., Aizawa, S., Park, H.S., Matsuda, T., Harada, K., Shindo, D., 2014. Advanced split-illumination electron holography without Fresnel fringes. Ultramicroscopy 137, 7–11.

Todokoro, H., Saitou, N., Yamamoto, S., 1982. Role of ion bombardment in field emission current instability. Jpn. J. Appl. Phys. 21 (10R), 1513.

Tonomura, A., 1973. Optimum design of accelerating electrodes for a field emission electron gun. Jpn. J. Appl. Phys. 12 (7), 1065–1069.

Tonomura, A., 1984. Applications of electron holography using a field-emission electron microscope. J. Electron Microsc. 33, 101–115.

Tonomura, A., 2011. Electron Holography, new ed. Springer, Berlin. OCLC: 751531756.

Tonomura, A., Matsuda, T., Endo, J., 1979a. Spherical-aberration correction of an electron lens by holography. Jpn. J. Appl. Phys. 18 (7), 1373–1377.

Tonomura, A., Matsuda, T., Endo, J., Todokoro, H., Komoda., 1979b. Development of a field emission electron microscope. J. Electron Microsc. 28, 1–11.

Tonomura, A., Osakabe, N., Matsuda, T., Kawasaki, T., Endo, J., Yano, S., Yamada, H., 1986. Evidence for Aharonov-Bohm effect with magnetic field completely shielded from electron wave. Phys. Rev. Lett. 56 (8), 792–795.

Troyon, M., 1980. High current efficiency field emission gun system incorporating a preaccelerator magnetic lens. Its use in CTEM. Optik 57, 401.

Uhlemann, S., Haider, M., 1998. Residual wave aberrations in the first spherical aberration corrected transmission electron microscope. Ultramicroscopy 72 (3–4), 109–119.

Van der Heide, P., 2014. Secondary Ion Mass Spectrometry: An Introduction to Principles and Practices. Wiley, Hoboken, NJ.

Vaz, C.A.F., Bland, J.A.C., Lauhoff, G., 2008. Magnetism in ultrathin film structures. Rep. Prog. Phys. 71 (5), 056501.

Veneklasen, L.H., Siegel, B.M., 1972. A field-emission illumination system using a new optical configuration. J. Appl. Phys. 43 (12), 4989–4996.

Völkl, E., Allard, L.F., Joy, D.C., 1999. Introduction to Electron Holography. Springer, New York. OCLC: 887208334.

Welford, W.T., 1976. Aplanatism and isoplanatism. Prog. Opt. 13, 267–292.

Welford, W.T., 1989. Aberrations of Optical Systems. Adam Hilger, Bristol, Philadelphia.

Williams, D.B., Barry Carter, C., 2008. Transmission Electron Microscopy: A Textbook for Materials Science, second ed. Springer, New York.

Wollnik, H., 2014. Optics of Charged Particles. Academic Press, Orlando. OCLC: 1044728471.

Xu, N.S., Huq, S.E., 2005. Novel cold cathode materials and applications. Mater. Sci. Eng. R Rep. 48 (2–5), 47–189.

Yakovlev, S., Libera, M., 2008. Dose-limited spectroscopic imaging of soft materials by low-loss EELS in the scanning transmission electron microscope. Micron 39 (6), 734–740.

Yao, N. (Ed.), 2007. Focused Ion Beam Systems: Basics and Applications. Cambridge University Press.

Yavor, M., 2009. Optics of charged particle analyzers. Adv. Imaging Electron Phys. 157.

Yavor, S.Y., Dymnikov, A.D., Ovsyannikova, L.P., 1964. Achromatic quadrupole lenses. Nucl. Instrum. Methods 26, 13–17.

Yurtsever, A., Zewail, A.H., 2012. Direct visualization of near-fields in nanoplasmonics and nanophotonics. Nano Lett. 12 (6), 3334–3338.

Zernike, F., 1938. The concept of degree of coherence and its application to optical problems. Physica 5 (8), 785–795.

Zewail, A.H., 2006. 4D ultrafast electron diffraction, crystallography, and microscopy. Annu. Rev. Phys. Chem. 57, 65–103.

Zewail, A.H., Thomas, J.M., 2010. 4D Electron Microscopy: Imaging in Space and Time. Imperial College Press; Distributed by World Scientific Pub, London, Hackensack, NJ. OCLC: ocn299716296.

Zhu, X., 1989. A computer program for electron gun design using second-order finite elements. J. Vacuum Sci. Technol. B: Microelectron. Nanometer Struct. 7 (6), 1862.

Zhu, X., Munro, E., 1995. Second-order finite element method and its practical application in charged particle optics. J. Microsc. (Oxford) 179 (2), 170–180.

Zienkiewicz, O.C., Morgan, K., 2006. Finite Elements and Approximation. Dover Publications, Mineola, NY. OCLC: ocm69792706.

Zuo, J.M., 1992. Automated lattice parameter measurement from HOLZ lines and their use for the measurement of oxygen content in $YBa_2Cu_3O_7$ from nanometer-sized region. Ultramicroscopy 41 (1–3), 211–223.

Zuo, J.M., 2017. Advanced Transmission Electron Microscopy: Imaging and Diffraction in Nanoscience. Springer, New York, NY.

Index

Note: Page numbers followed by "*f*" indicate figures and "*t*" indicate tables.

Kolibri 360 GmbH
Ferdinand-Jühlke-Straße 7
99095 Erfurt, Deutschland
produktsicherheit@kolibri360.de